Roman Amphorae from the Amber Route in Western Pannonia

Tamás Bezeczky

BAR International Series 386
1987

B.A.R.

5, Centremead, Osney Mead, Oxford OX2 0DQ, England.

BAR S386, 1987: 'Roman Amphorae from the Amber Route in Western Pannonia'

ISBN 9780860544999 paperback

ISBN 9781407346632 e-book

DOI https://doi.org/10.30861/9780860544999

A catalogue record for this book is available from the British Library

This book is available at www.barpublishing.com

List of Contents

List of Plates

List of Tables

Mineral analyses

List of Figures

Figure		Page

The numbers of the sites

1. Aguntum
2. Aisilingen
3. Aquileia
4. Aquincum
5. Asolo
6. Asti
7. Aventicum
8. Baone/Este/
9. Bergamo
10. Carnuntum
11. Carpi
12. Casteggio
13. Cecina
14. Celeia
15. Chiavenna
16. Chur
17. Como
18. Concordia
19. Cortaccia/Alto Adige/
20. Cremona
21. Emona
22. Ercsi
23. Fažana/Fasana/
24. Feltre
25. Fornovo d'Adda
26. Gomolava
27. Gorsium
28. Intercisa
29. Loron-Poreč (Parentium)
30. Magdalensberg - Helenberg
31. Milano
32. Monteu da Po
33. Mortara
34. Mursella
35. Nimes
36. Oberstimm
37. Ovilava/Wels/
38. Padova
39. Paperiano
40. Parma
41. Poetovio
42. Porolissum
43. Portorecanati
44. Pula/Pola/
45. Reggio Emilia
46. Rimini
47. Roma
48. Rovinj
49. Salla
50. Savaria
51. Spinimbecco/Legnago/
52. Teutoburgium
53. Tortona
54. Trieste
55. Vallera
56. Valeia
57. Vercelli
58. Vienne
59. Vincenza
60. Virunum
61. Visinada
62. Vindobona

INTRODUCTION

The Amber Route is one of the ancient roads of commerce of our continent. Works of fundamental importance have been published about the history of the population living here and settlements situated along the road.[1] The inscribed stone relics encountered here, as well as the terra sigillata, glass and other finds have already been described, but the amphorae - of equal significance - have not yet been published,[2] although by doing so further information could be gained about the commercial connections of the Roman Empire.

By launching this discussion of hitherto neglected finds, the present paper seeks to contribute to a more detailed knowledge of the history of the province of Pannonia.

Amphora research is traditionally based on the investigation of the form, inscriptions and stamps. These considerations are supplemented by scientific (petrological and neutron activation) analysis of amphorae fabrics.

By establishing the origins of amphorae, the trade in some important foodstuffs (wine, oil, olive, fish sauce etc.) may be determined as well, for the places of production of these were identical with the places of manufacture of the amphorae in which they were stored. At the same time amphorae are relatively easy to date, thus enabling the dating of other finds of the sites as well.

In describing the finds, the classical methods of amphorology have been utilized.

H. Dressel established the connection between the various types of the contents of amphorae.[3] He imposed a table of forms upon the finds and this classification has been used by many archaeologists - with certain supplements and changes[4] - up to the present day. In analyzing the finds on the Amber Route, one of the points of reference are those publications of P. Baldacci in which Cisalpine amphora types are evaluated.[5] Baldacci described the Dressel 6 type in full detail; it is the most common single form on the Amber Route. For the same type E. Buchi[6] distinguished two sub-types, Dressel 6A and 6B.

In describing the individual forms, I have used the method of E. Hamon and A. Hesnard.[7] (This formalized method of description was utilized in the computer analysis classifying the individual types on a mathematical-statistical basis, see supplement 2.)[8]

The petrological analysis[9] determined the composition of the rocks and minerals added to the clay of the amphorae. The metals occuring as trace elements in the clay were indicated by neutron activation analysis.[10] These important analyses could only be performed in a limited number on account of the meagre financial resources available. However, whenever we managed to obtain several samples from the clay of one workshop, the analysis were performed on all of them.

The finds were collected from the following sites: Nauportus (Vrhnika), Emona (Ljubljana), Celeia (Celje), Poetovio (Ptuj), Salla

(Zalalövő), Savaria (Szombathely), Scarbantia (Sopron), Carnuntum (Petronell, Deutsch-Altenburg).[11]

The amphorae from Aquileia will be published soon by a team of researchers from Rome and Trieste Universities, coordinated by M.B. Carre, a member of the Ecole Française de Rome.[12]

In the present study, we wish to describe primarily the finds made in the vicinity of the Amber Route but reference will be made in addition to finds from Vetus Salina, Gorsium and Aquincum in eastern Pannonia.

In the drawn and photographed tables the finds have been indicated in each case by the numbers contained in the catalogue.[13]

ACKNOWLEDGEMENTS

The work of collecting and processing the finds discussed in this study was assisted by Professor István Fenyő (Department of Mathematics, University of Technology, Budapest) and Professor András Mócsy (Department of Archaeology, Eötvös University, Budapest) within the state commision received from the Ministry of Education in the years 1981-85.

Efforts to get acquainted with the finds and the circumstances were assisted by a number of experts and museums.

I extend my heartfelt gratitude for the assistance rendered my work by Éva Bónis, Teréz Buócz, Vajk Cserményi, Jenő Fitz, János Gömöri, László Horváth, László Kocsis, Magdolna Medgyes, András Mócsy, Ferenc Redő, Klára Póczy, Tihamér Szentléleky, Eszter Szőnyi, Endre Tóth and Judit Topál.

When collecting material in Jugoslavia, I received valuable help from Blagoj Jevremov, Vladimir Kovačič, Sonja Petru, Ljudmilla Plesničar-Gec, Branka Šulc and Elica Boltin-Tome. I investigated the Aquileian amphorae with the kind permission of Luisa Bertacchi, and the Triestian amphorae with that of Franca Scotti Maselli. I am grateful to Gino Bandelli, Maria Teresa Cipriano, Marco Fano and Clementina Panella for helping me to get acquainted with finds in Italy.

I wish to thank László Barkóczi, András Mócsy, Miklós Szabó, János György Szilágyi, Endre Tóth and Marie Brigitte Carre for their valuable advice and critical remarks and Paul R. Sealey for correcting the English version.

The Amphora types along the Amber Route

I. DRESSEL 2-4 (Nos. 1 - 43)

The characteristic features of these related forms are the bead rim and the cylindrical neck. There is a sharp break at the junction of the neck and the body; the body has an arched profile. The handles are formed from two rods; the whole form has components which are easy to distinguish. Amphorae of this type were used for wine.

Among the finds along the Amber Road, an amphora could be identified from Tarraconenis (No. 2) on the basis of its characteristic form.[14] The place of origin of the other pieces is difficult to define on the basis of their form. The following groups could be distinguished. Identical formal characteristics are possessed by Nos. 6 and 7 from Poetovio, Nos. 23, 26, 27, 30 and 32 from Savaria, as well as No. 13 from Salla; these amphorae are similar in colour and fabric as well. Other properties are characteristic of finds from Salla, Nos. 15, 16, 17, in which the handles were not made from two individual parts, the division being indicated only by a groove.

Fabric

Several adjacent workshops may be called a "site of production". Though they can be readily distinguished, it is impossible - for the time being - to determine the sites of production of the pieces which contain intermediary mineral and rock granules. These granules are the augite (No. 19) and the sanidine (No. 17). With volcanic raw materials in the clay, the sites of production must have been somwhere in Central and Southern Italy and Southern Gaul. The composition of the fabric of two amphorae (No. 7 and No. 31) is very similar to that of the pieces made in the workshop of C. Laecanius Bassus (see below, Tables 4-5) [FELIX.SR (Aquileia 376), HOM (Aquileia 373), VIAT (Aquileia 378), VIAT (Aquileia 443), CLYMEN (Poetovio 10094).]

Chronology

A handle fragment from Salla (No. 13) was brought to the surface from layer 5, while the sherd from Salla (No. 21) dates to the second period of the military camp. Accordingly, the fragment No. 13 may be dated A.D. 85-120 and No. 21 A.D. 50-85.[15] Nos. 28 and 31 from Savaria have been found in the layer dated to the second hald of the 1st century A.D.,[16] the fragment No. 33 was used for covering an urn grave in the 1st century A.D.[17]

There is no data available to date the other finds.

II. DRESSEL 5 (Nos. 44 - 45)

A specimen of this type was found in Poetovio (No. 44) and another in Savaria (No. 45). On the Poetovio amphora we find letters painted in red, but barely legible: P..T (Pl. 1). One of these amphorae from the Magdalensberg has a Greek inscription. This may confirm Zevi's view that these forms come from the Greek Islands.

Fabric

The analyses have demonstrated (Table 1 and 2) that both amphorae contain a low proportion of quartz (12%), augite (2.1 - 1.9%), quartzite (2.3 - 2.8%), traces of plagioclase and significant proportions of micrite (59.8 - 79.9%). Here augite indicates the place of preparation to be of volcanic origin.[18] In the fabric of amphora No. 34 clastic andesite could also be identified (6.9%).

III. DRESSEL 6 (Nos. 46 - 193)

The variants of Dressel 6 were clearly distinguished by Baldacci[19] when describing the types from Cisalpine and by Buchi[20] when publishing the types from Verona.

From Dressel 6A exists in several variants and it is partly identical with the form called II.a. by Baldacci.

A few amhoprae of this type are known from Emona[21] along the Amber Route (Nos. 46-48). Their formal characteristic is that the handles are curved, joining the upper part of the body in an arch; the body has an oval shape, ending in a long spike at the bottom. The rim joins the long arched neck with a sharp protrusion.

Dressel 6B (Baldacci III.a. and II.b.) is characterized by a rim shaped like a chalice, joined to the neck without a transition; the profile is continuous along the line of the handles and the body.

The body is pear-shaped or oval, mostly widening towards the bottom and ending in a knob.

Several sherds of this type have been found at sites along the Amber Road; the overwhelming majority of them were manufactured by C. Laecanius Bassus, Calvia Crispinilla and then, from the last quarter of the 1st century A.D., by imperial workshops. In the following these workshops will be analyzed in detail.

III.1. The workshop of C. Laecanius Bassus

We know that the Laecanius gens[22] was Italian, probably Etruscan in origin; its members were among those founders of the Pola colonia who were adopted into the aristocracy after the civil war.

The first well-known representative of the family is C. Laecanius Bassus[23], who became praetor urbanus in A.D. 32, consul suffectus in A.D. 40, and later (A.D. 64) his son of the same name held the same office.[24]

The existence of his estate in Istria is proved by the Matterian inscription[25] from the period of Claudius. He had high quality lands on the territory of Pola colonia too, on one of them may have been located the pottery of Fažana, where stamps of Laecanius[26] were found on masses of amphorae. A. Gnirs excavated a number of amphorae with their rims turned downwards in the store room of a villa,[27] in the Val

	Stamp		Findspot
42.	IMP.AVG.GER		Emona No. 70.
			Poetovio No. 83.
			Poetovio CIL III. 14112.1.
			Salla No. 102.
	IMPDOMITI .E.SR		Savaria No. 154.
43.	IMP.NERVAE.AVG		Poetovio No. 84.
52.	CAL.CRISPINILLAE		Emona No. 55.
			Emona No. 56.
			Poetovio No. 77.
			Poetovio No. 79.
			CIL III. 12010.2.
			Carnuntum No. 190.
	CRISPINILI		Poetovio No. 80.
			Poetovio No. 81.
54.	C. Laecanius Bassus		
	CLA .BASS		Poetovio CIL III. 14371.4.
	LAEI		Carnuntum CIL III. 12010.20.
			Salla No. 96.
	LA B		Poetovio No. 74.
	C.A.K...		Emona No. 53.
	C.LAEK BASS	AMETHYSTI	Poetovio CIL III. 14371.5.
	C.LAEK B	BAR	Poetovio CIL III. 12010.3.
	C.LAEK BAS	CLYMEN	Poetovio No. 76.
	CLAEK	FI\	Emona No. 54.
	C.LAEK.BAS	SPERATVS	Emona No. 51.
	IAEk	VIAT	Emona No. 52.
			Savaria No. 150.
		FELIX	Poetovio CIL III. 12010.16.
		FEX.PE	Salla No. 97.
		FELC	Poetovio No. 75.
		FEL.TVR	Salla No. 107.
55.	COSAE		Emona No. 60.
56.	T.FLAV.FON		Emona No. 68.
58.	APICI		Emona No. 61.
59.	L.IVN.PATIEN		Emona No. 66.
			Emona No. 67.
69.	/VARI PA/CCI		Emona No. 63.

Figure 1 Dressel 6B stamps on Amber Route amphorae, listed by Baldacci number

San Pietro bay of Pola. On the island of Brioni too, large quantities of amphorae were found in the villa of Val Catena[28], also with stamps of Laecanius.

In connection with the workshop at Fažana, it has been mooted[29] whether or not we can be sure its owner was indeed Laecanius Bassus. It could conceivably have belonged to an independent entrepreneur who made amphorae for him. After the name of the senator, the names of the heads of the workshops frequently appear as well. They always pressed their stamps on the rim above one of the handles; that of Laecanius was always on the rim between the handles. This regularity can also be observed on the amphorae along the Amber Road as well.

From a compilation published by F. Tassaux[30], we know the names of some thirty vilici, (i.e. workshop heads) of Laecanius (Fig. 2). Among them, the following names have been published from the Magdalensberg[31]: ADEL(phus), BARB(...), BARD(...), COM(...), EVHAR(istus), FELIX, FELIX.PET, FELIX.SCR, FELIX.SER, FEL.TVR, HER(...), SPERATVS, VRBA(nus), and Viat(or). These masters were active prior to A.D. 45; the amphorae were manufactured before the site ceased to operate. We do not know when precisely amphorae made in the workshop of Laecanius were produced. But at least a fragment with a stamp on it - LAE - was excavated at the Magdalensberg[32] from a layer that dates A.D. 15 - 20. It was about this time that the workshop may have begun to operate and export its products.

From the inscriptions of the stamps published to date, it is apparent how the individual stamps changed and were transformed:

C.LAEK.BASSI; CAE B; LAE ; LAEk.

It may also be observed that the stamp changed in the case of the same vilicus or workshop head as well

LAEk; VIAT Emona (No. 52), Savaria (No. 150)

LAEk; VIAT Aquileia (No. 443).

In the case of the pieces from Emona (No. 52) and Savaria (No. 150), stamped with the same die, significant differences may be observed in the moulding of the form. In the case of ceramic containers of such size, naturally, there may be differences[33] because amphorae were not assembled from one single piece. It is also possible that the component parts were made at different times.

Among the amphora stamps from the sites along the Amber Route (Fig. 3), the stamp C.LAEk.BAS(SI);SPERATVS appears only in Aquileia and Emona (No. 51). Moreover, in Emona, in addition to the stamp VIAT/or/ (No. 52), on a stamp difficult to read with certainty, the name FELIX (No. 54) appears as well.

At Poetovio the names of AMETHYST (CIL III. 14371.5), BAR/.../ (CIL III. 12010.3.) and FELIX (CIL III. 12012.16) are mentioned. In addition, a stamp of FEL/ix/ Of/icina/[34] (No. 75) has also been published. The reading is not unequivocal. The stamp C.LAEkBAS;

Adel(phus) : Aquilée, PAIS, 1077,20 = BALDACCI, 36 (seul); Magdalensberg, *Ausgrabungen*, 1950, p. 86 n° 1.

Amethystus : Fasana, *JA.*, p. 84 n° 1 et fig. 5-6. *Poetovio*, *J. Alt.*, p. 84 fig. 4.

Amycus : Val San Pietro, *JÖAI*, 1910, p. 102, 18 cas. Pola, *JA*, p. 86.

Barb(ius) : Fasana, *JA*, p. 84 n° 5; *JÖAI*, 1910, p. 97; *JÖAI*, 1911, p. 3 deux cas. Val Catena, *JA*, p. 84 n° 5 et fig. 9; *JÖAI*, 1904, p. 145. Poetovio, *CIL*, III 12010,1 et 3. Magdalensberg, *Ausgrabungen*, 1949, p. 486 n° 4; 1958, p. 180 n° 23; 1965, p. 367 n° 15 (ou *Bard?*); 1970, p. 299 n° 2 (seul).

Car(us) : Aquilée, PAIS, 1077, 89.

Clarus : Rome, *CIL*, XV 3477 = CALLENDER n° 365.

Colon(us) : Fasana, *JA.*, p. 85 n° 11; *JÖAI*, 1910 p. 97 (seul), même timbre et amphore que ceux de *Laek.*; *JÖAI*, 1911, p. 38 deux cas.

Clymen(us) : Verceil, *CIL*, V 8112, 52 = CALLENDER n° 365 = BALDACCI, p. 36 Mortara, *Boll. Soc. Pavese*, 1964, p. 150 = BALDACCI p. 36 (seul).

Com. ou Comi. : Fasana, *JA*, p. 85 fig. 6; *JÖAI*, 1911, p. 38 huit cas. Val San Pietro, *JÖAI*, 1910, p. 97. Val Catena, *JÖAI*, 1904, p. 145; *JÖAI*, 1906, p. 40; *JA.*, p. 85, fig. 9. Milan, A. FROVA, *Epigraphica* 14, 1952, p. 29 = BALDACCI p. 00. Magdalensberg, *Ausgrabungen*, p. 180 n° 24 [12].

Crescens : Fasana, *JA*, p. 84 fig. 5-6; *JÖAI*, 1910, p. 97; *JÖAI*, 1911, p. 00.

Datia(nus) : Fasana, *JÖAI*, 1910, p. 97 et fig. 27.

Di . . . : Fasana, *JA*, p. 85 fig. 6; *JÖAI*, 1910, p. 97.

Euchar(istus) : Val San Pietro, *JÖAI*, 1910, p. 102 quarante-six cas. Aquilée, PAIS, 1077, 66 = BALDACCI p. 36 (seul). Magdalensberg, *Corinthia*, 1, 1958, p. 172 = BALDACCI p. 36.

Felix ser(vus) : Fasana, *JA*, p. 84-85. Brioni, *JA*, p. 86. Val San Pietro, *JÖAI*, 1910, p. 102, trente-deux cas. Aquilée, PAIS, 1077, 88, b et c; PAIS, 70 a = BALDACCI, p. 00 (seul, *Felix Pe*). Concordia, PAIS, 1077, 70 b = BALDACCI, p. 36. Padoue, *NS*, 1881, p. 240 n° 6 = PAIS, 1077, 88 a; PAIS, 1077, 71 = BALDACCI, p. 36 (seul). Magdalensberg, *CIL*, III 6007, 5; *Ausgrabungen*, 1949, p. 487 n° 6 (*Felix Pe*); 1950, p. 86 n° 6 (*Felix Pe*); 1951, p. 70 n° 6; 1959, p. 72 n° 5; 1962, p. 454 n° 5 (*Felix Pet*); 1970, p. 303 n° 1 (*Felix Scr.*) (seul); p. 315 n° 7. *Poetovio*, ?, CALLENDER, n° 623 = BALDACCI, p. 36 (*Fel. of*).

Her. ou Heri : Val Catena, *JA*, p. 86 fig. 9; *JÖAI*, 1908, p. 178 (Her). Val San Pietro, *JÖAI*, 1910, p. 102 (Heri) douze cas. Aquilée, PAIS, 1077, 11 (Her). Verceil, *CIL*, V 8112, 53 = CALLENDER n° 365 (Herme). Tortona, *NS* 1897, p. 373 = BALDACCI, p. 36 (Herne). Turin, B. S. Piémont, A 1918 p. 171 = BALDACCI, p. 36 (seul). Magdalensberg, *Ausgrabungen*, 1951, p. 70 n° 5; 1956, p. 152 n° 12; 1970, p. 301 n° 7 (Her). Il est vraisemblable que l'on ait ici deux personnages différents Hermes et Heri(anus).

Hom(o?) : Aquilée, PAIS, 1077, 90.

Ialys(us) : Fasana, *JÖAI*, 1911, p. 38 huit cas. Val Catena, *JÖAI*, 1904, p. 145 = J. Alt., p. 86 et fig. 9. Padoue, *NS*, 1881, p. 240.

Isar (Isaurus?) : Padoue, *NS*, 1881, p. 240 = PAIS 1077, 91.

Martius? : Il n'est cité que par A. GNIRS, *JA*, p. 86, sans précision du lieu de trouvaille, ni dessin.

Op(ius) : Fasana, *JA*, p. 85 et fig. 6; *JÖAI*, 1910, p. 97 (seul, mais «timbre et amphore identiques à ceux de Laecanius»).

Pier(is) : Fasana, *JA*, p. 84 et fig. 6; *JÖAI*, 1910, p. 97; *JÖAI*, 1911 p. 38, dix-sept cas. Padoue, *NS*, 1881, p. 240 = PAIS 1077, 92.

Rom(anus) : Fasana, *JA*, p. 85 (seul, même remarque que pour Opius et Colonus).

Sperat(us) : Aquilée, PAIS, 1077, 93; PAIS, 137 = BALDACCI, p. 36 (seul). Magdalensberg, *Ausgrabungen*, 1950, p. 86 n° 5; *Carinthia*, 1, 1959, p. 129 n° 17 et 1966, p. 454 n° 11, 12 = BALDACCI, p. 36.

Syntrophus : Verceil, *CIL*, 8112, 54 = CALLENDER, n° 366. Tortona, *NS*, 1897, p. 374 = BALDACCI, p. 36 (seul).

Urba(nus) : Magdalensberg, *Ausgrabungen*, 1951 p. 72 n° 7, 8, 9.

Viat(or) : Fasana, *JA*, p. 85 et fig. 6; *JÖAI*, 1910, p. 97; *JÖAI*, 1911, p. 38 trente et un cas. Val Catena, *JÖAI*, 1904, p. 145; *JÖAI*, 1906 p. 40 = *JA*, fig. 00. Aquilée, PAIS, 1077, 152 = BALDACCI, p. 36 (seul). Padoue, *NS* 1931, p. 158 = BALDACCI, p. 36. Crémone, BALDACCI, p. 36 = Musée, Cat. Pontiroli. Magdalensberg, *Carinthia*, 1, 1961, n° 41 = BALDACCI, p. 36.

. . . es : Fasana, *J. Alt.*, p. 85 fig. 6; *JÖAI*, 1910, p. 97.

Figure 2 List of vilici or freedmen of Laecanius' workshop compiled by Tassaux

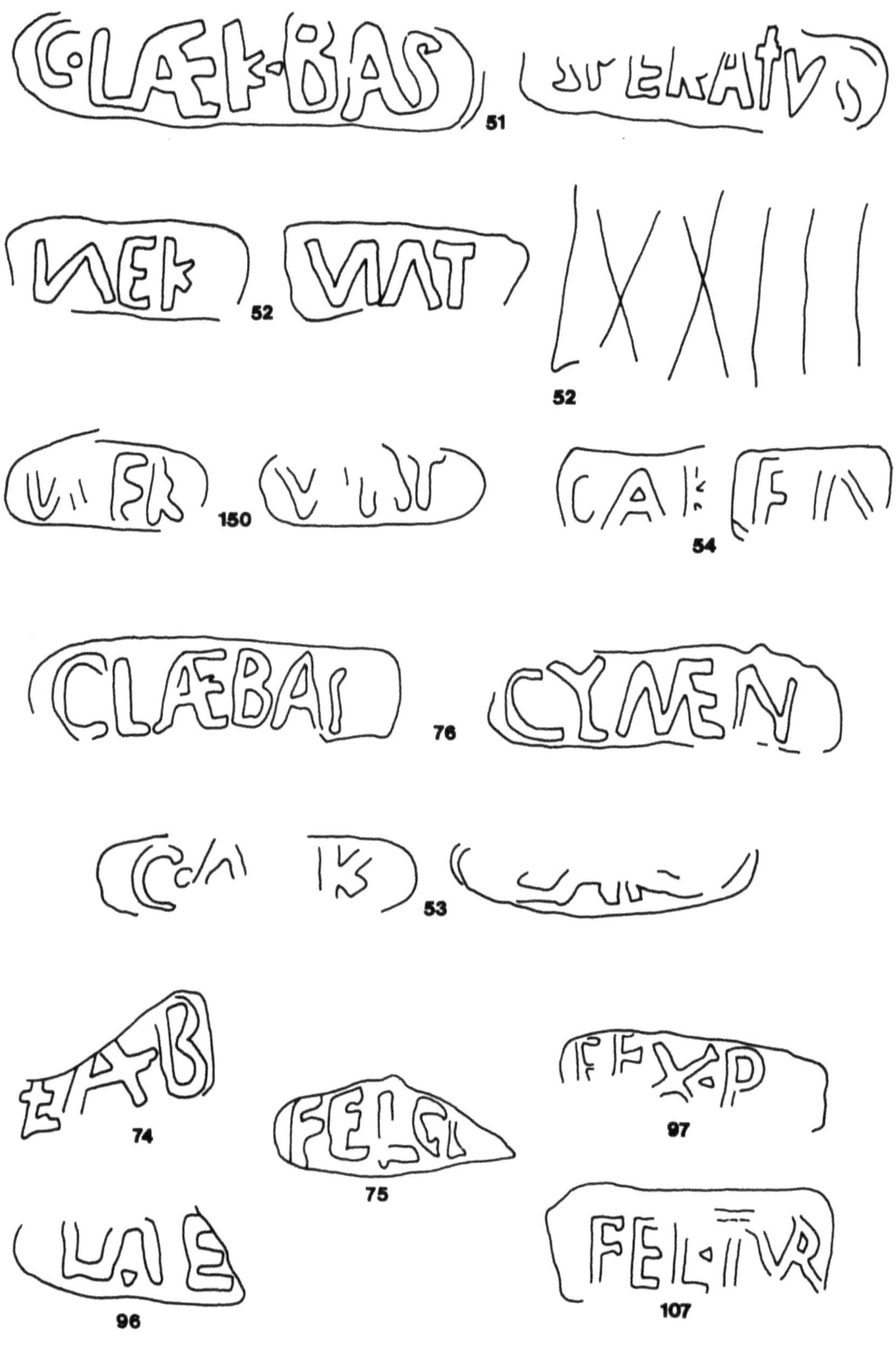

Figure 3 Stamps of Laecanius along The Amber Route

CLYMEN(us) (No. 76) is also attested at Poetovio. In connection with a similar stamp, Baldacci[35] supposes that following the death of Laecanius, the workshop was taken over by the Emperor.

He has demonstrated that the name of CLYMEN(us) occurs with that of Laecanius on one occasion, and on another with a stamp IMP (CIL V. 8112.6.). In the workshop the forms continued to be manufactured after the owner had changed too, with certain minor modifications. Even the stamps of Hadrianus[36] appeared on amphorae of this type, but later manufacture was probably limited to local needs.

We have the stamps FEX.PE (No. 97), LAE (No. 96) and FEL.TVR (No. 107) from Salla and LAEK, VIAT(or) (No. 150) from Savaria.

At Carnuntum the stamp LAE (CIL III. 12010.20.) has come to light. Amphorae with the stamp of C. Laecanius Bassus are not known at Celeia and Scarbantia. Naturally, in Aquileia several amphorae are to be found from the Fazana workshop[37].

The petrological analysis of the amphorae of Laecanius

The quality of the fabric used in the workshop of Laecanius changed and the temperature of burning was not always the same either. Tables 4 and 5 demonstrate the results of geological investigations. The pieces collected in Aquileia were also analysed. The result showed that some amphorae produced in the workshop are basically different from the majority. One must ask if these sherds were made of the same clay. The amphorae made in this workshop are characterized by a large quantities of carbonated (micrite, sparite). Sometimes large numbers of fossils (foraminiferan, mollusc, spicula) occur from marine clay. The material of the amphorae contains very little magma granule, but in some cases larger amounts of phillite can be identified (No. 96).

In the case of one piece (not found along the Amber Route) we had the opportunity to take samples from the rim, the neck, the lower part of the body and the handle. The amphora was found at Aquincum[38], with the stamps CLAEKB; BAR (Pl. 13, Table 3).

From the geological analysis, it emerged that several fabrics were used when making the amphorae. The element that played a part in lifting and transportation, the handle, was made from tougher material than the rest. It has a higher quartz, biotite and muscovite content than the rest and lacks carbonate.

But neutron-activation analysis proved beyond doubt that the basic material used originated in the same place. While making the amphorae, fabrics appropriate for the individual functions were used for the different parts. That means that the production was well organized and careful attention was paid to technological requirements.

The dating of the amphorae of Laecanius

The fact that the amphorae made in the workshop can be well dated is important from the point of view of investigations in Pannonia since the amphorae of Laecanius may be found not only along the Amber Route but on other sites in our province too.[39]

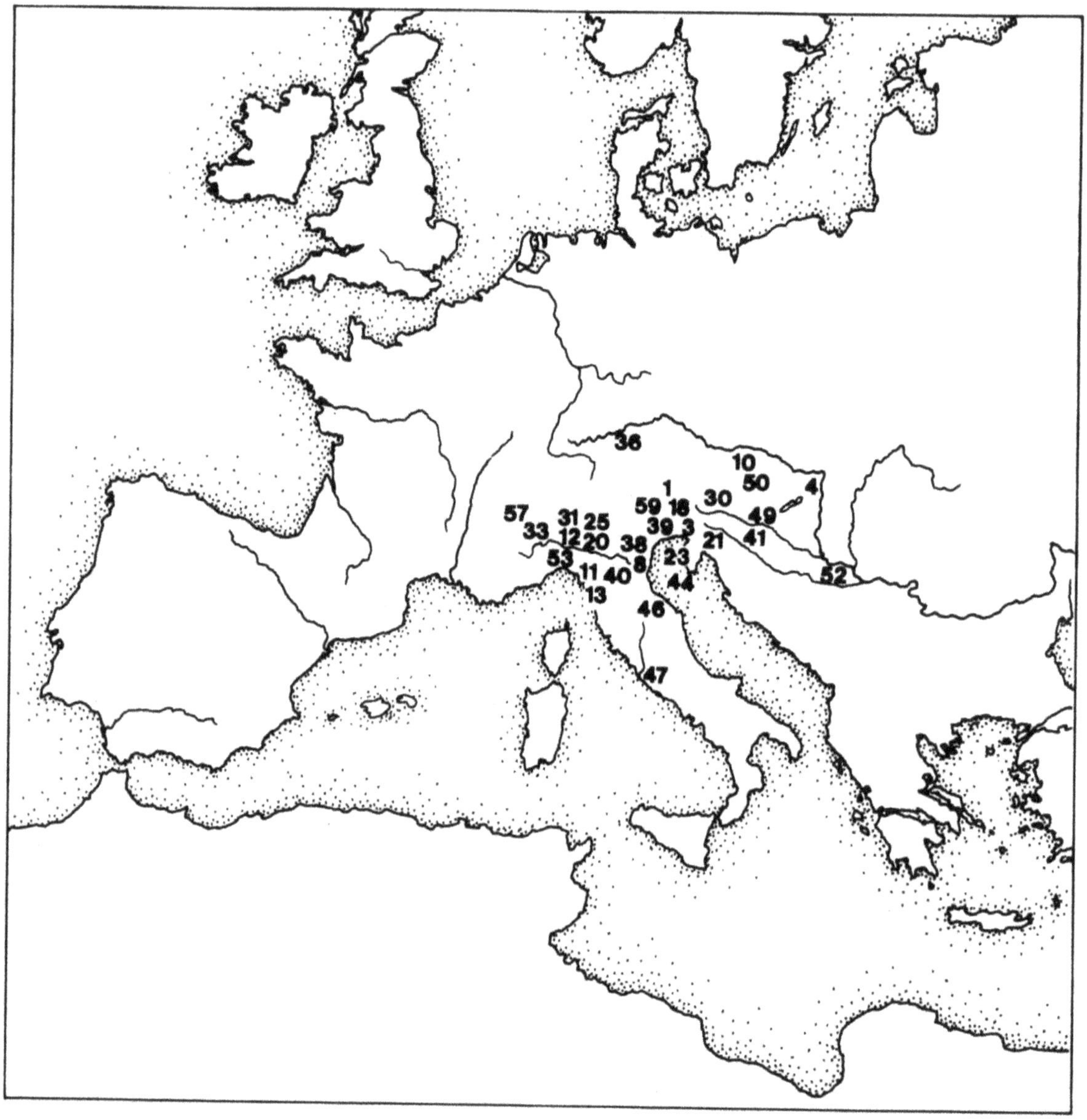

Figure 4 Distribution of Laecanius' amphorae

According to our present knowledge, the amphora with the stamp LAE, from the Magdalensberg, dated between A.D. 15-20, may indicate the beginning of the operation of the workshop. Presumably Laecanius Bassus retained the workshop until his demise. According to the investigations of Tassaux, when Laecanius senior became praetor urbanus A.D. 30, he was at least 27 years old.[40] A. Degrassi reconstructed[41] the name of C. Laecanius Bassus P(ater) in the Soldales Augustales Claudiales list of names discovered not long ago, dated 64, when Laecanius junior was consul with M. Licinius Crassus Frugi. Apart from being a consul, all we know about Laecanius junior from Plinius[42] is that he died shortly before 78. In this connection Tassaux mentions[43] that the senatorial branch of the Laecanius Bassus family was jeopardized by dying out; therefore it too, like other senatorial families, tried to save itself by adopting children.

If that is true, then we feel it is realistic to propose the year A.D. 80 (suggested by M.H. Callender) as the time when the workshop passed over to the hands of the emperor's house.[44]

The distribution of the amphorae made in the workshop of Laecanius

The distribution of the amphorae of Laecanius is illustrated by Figure 4. As compared to the sites known so far,[45] the picture formed before is supplemented by the stamps found at Emona, Salla, Savaria, Aquincum and Teutoburgium. It can be expected that amphorae from this workshop will eventually be noted at other sites in Pannonia as well.

III.2. The workshop of Calvia Crispinilla

The other well-known factory owner was Calvia Crispinilla, at whose workshop several amphorae of type Dressel 6B were manufactured.

Tacitus[46] refers to him as "Magistra libidinum Neronis", then remarks that at the time of Nero's fall, Crispinilla had to leave Rome at once because the people grew angry with him because of his misdeeds. Later on, however, he regained the respect of the whole City. By way of his marriage of consular rank, he made his position secure and then his wealth made him an influential person. He marked amphorae with the stamps CAL(via) CRISPINILLAE and CRISPINILI (Fig. 5). We also know stamps TRAVL ET CRISP and TAFCRISPINAE, but in connection with these, Tassaux[47] has established that they are not related to the workshop of our Crispinilla. It has frequently happened that the stamp CRISPINILI is read CRISPINI O(fficina) (Plate 5). This was so in the case of M. Abramic[48] too, whose faulty reading was taken over by Callender[49], Baldacci[50] and Tassaux.[51] This may have happened because of the incomplete documentation; no photograph was published of the stamp. The stamps published erroneously from Aislingen[52] and Oberstimm[53], have been corrected by E. Schallmayer.[54]

Among the finds along the Amber Road we encounter the amphorae at Emona (No. 55, 56), Poetovio (No. 77-79), Carnuntum (No. 190) and, naturally, Aquileia[55], but one marked with a stamp is also found at Mursella along the road from Savaria to Arrabona.[56]

The stamp CRI?PIN appears also on two amphorae of this type at

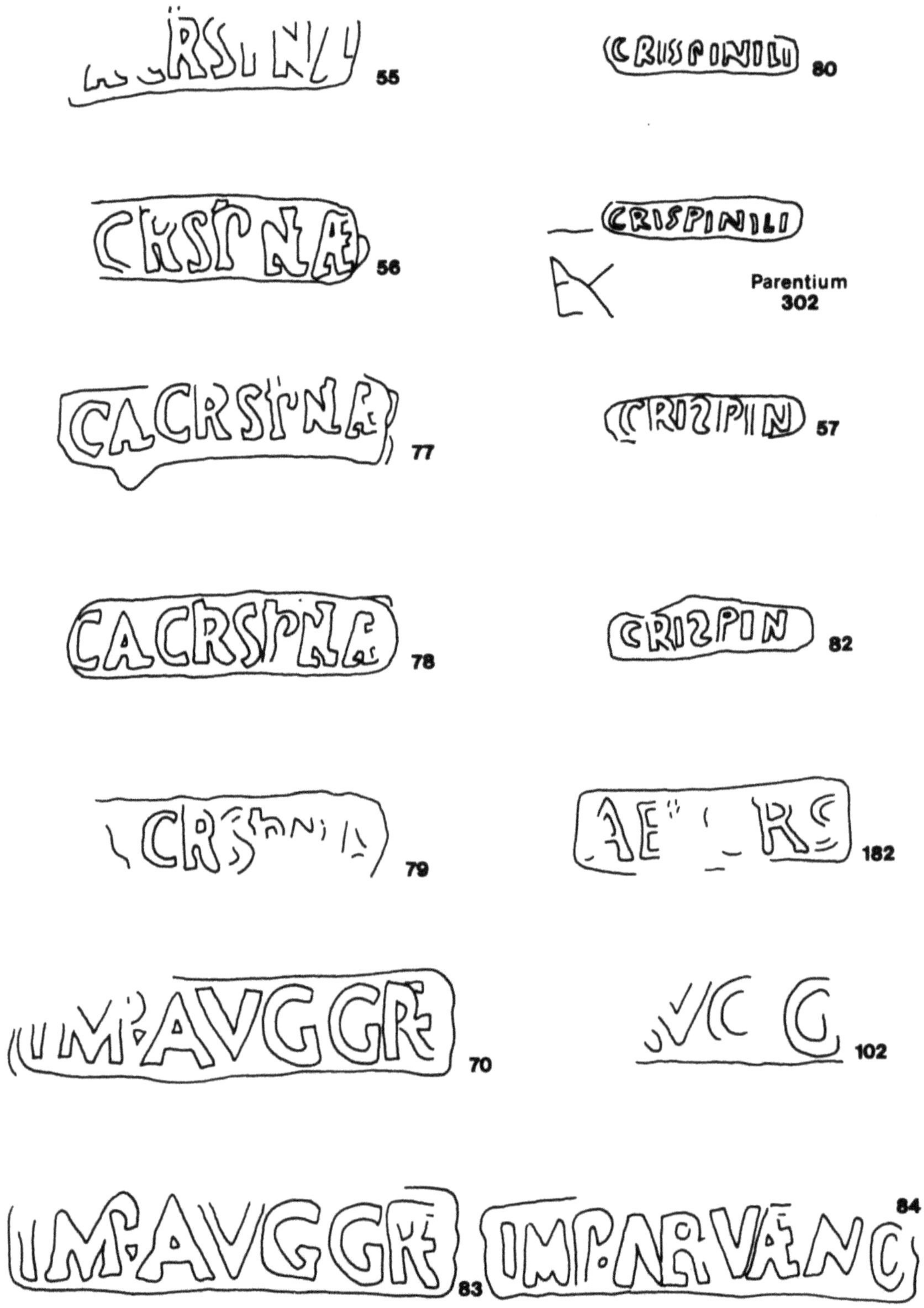

Figure 5 Stamps of Crispinilla and the Emperors along the Amber Route

Emona (No. 57) and Poetovio (No. 82). The place of manufacture of these pieces is not known, but we know that they are present among the finds at Aquileia[57] and the Magdalensberg[58]; therefore they may be dated in the first half of the first century A.D.

The workshop of Loron

It seems that the workshop of Calvia Crispinilla operated at Loron, near Porec.[59] During investigations conducted with V. Kovačič in 1984, we collected a number of fragments of amphorae (cf. Fig. 37), lucernae and glass along the seashore, in the vicinity of the 50-60cm wide base walls that have remained extant from the alleged workshop. From the sea and in the layers of the shore we found innumerable red pottery sherds.

During the last century amphora stamps with the names of C. Crispinilla, Domitianus, Nerva and Traianus were collected here.[60]

So far no excavations have been carried out on the site of the Loron workshop, therefore we do not know when it started to operate; we have no information either about when it was taken over by the emperor.

A fragment housed in the museum at Poreč, a rim, amplifies our knowledge.[61] Before firing, the stamp CRISPINILLI was put on the amphora, then after firing the signature of C. Laecanius Bassus was cut into it. If the sign originates from Laecanius (Pl. 3 lower picture), then what probably happened was that for some reason the ready-made amphorae of the neighbouring workshop were taken over. It is possible that there were not enough amphorae in the workshop at Fažana, or that after the stormy departure of C. Crispinilla, Laecanius made use of the ready-made pieces of the nearby workshop, or, perhaps, took over the workshop as well! Since oil produced by Laecanius was put in the amphora, his signature may have been added as a guarantee.

The stamp AELI.CRIS was also collected at Loron; it is also present at Poetovio (No. 94) and Savaria (No. 182).

The petrological analyses of the amphorae of Crispinilla

Relatively little additives were mixed with the fabric of the amphorae of C. Crispinilla (2.4 - 5.5%).

The data on the geological analyses is given in Tables 6 and 7.

In the case of pieces with the stamps CRISPINILI, the quantity of granules (1.1 - 5.4%), the proportion and composition of the individual rocks and minerals were similar to sherds bearing the stamp CAL.CRISPINILLAE.

The fabric of the amphorae marked with the stamp CRI?PIN did not have similar properties to those of the two groups described above.

The distribution of amphorae made in the workshop of C. Crispinilla

Emona, Poetovio, Carnuntum and Mursella can be added to the area dover which the amphorae with the stamp CAL.CRISPINILLAE are scattered.

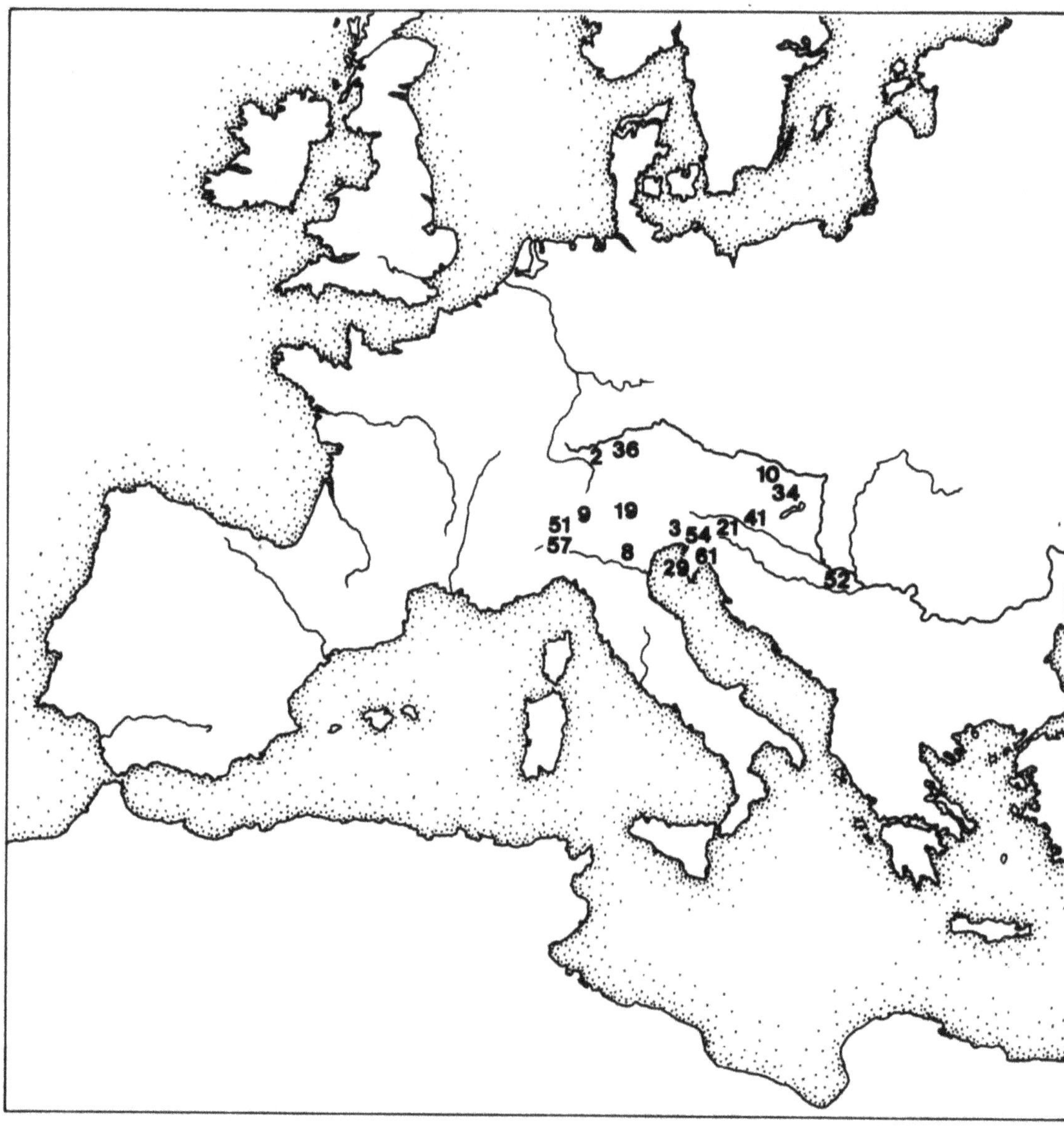

Figure 6 Distribution of Crispinilla's amphorae

The distribution of pieces with the stamp CRISPINILI is Aquileia, Poetovio, Aislingen and to these - already known - sites we may add Oberstimm, Teutoburgium and Porec.

When comparing the distribution of the amphorae made in the Crispinilla's and Laecanius' workshop, the latter of which was larger and had started to operate earlier, it emerges that Laecanius' products were spread over a wider area.

Their amphorae are found together in the sites of Cisalpina, Noricum and Pannonia, but the amphorae of Crispinilla are found in two Raetian camps as well (Aisilingen and Oberstimm).

III.3. Amphorae with imperial stamps

Baldacci classified amphorae with imperial stamps in his group II.b. and stated that they do not actually have any diagnostic shape[62]. From the formal-typological point of view, the doubts of Baldacci seem to be well-founded in connection with amphorae of Laecanius, Crispinilla and imperial workshops. The type Dressel 6B was established by the transformation of the classic form of Dressel 6 which became the best-known type in Cisalpina, Noricum and Pannonia from the first quarter of the first century A.D. Slightly modified, it remained in use until the end of the first quarter of the 2nd century.

Petrological analyses

The materials added to the clay of amphorae with IMP stamps (Table 8) naturally show a picture similar to that of the products of Laecanius and Crispinilla. Thus, perhaps, Baldacci's supposition may be justified that after the proprietor had changed, only the stamps were altered and production continued in an unchanged manner.

Most of the amphorae with imperial stamps may be classified in group V. of Laecanius' workshop (Aquileia 381, Aquincum 41673, Aquincum 55.304) while others clearly resemble group I. (Nos. 83, 102 and Porec 306) and group II. (Nos. 70 and Porec 304) of Crispinilla's workshop.

Unstamped dressel 6B amphorae

In the case of unstamped sherds in this group, it is impossible to decide on typological grounds which workshop produced them. Since in the majority of cases it is only fragments, all that may be stated is that their manufacturing method was very similar to the method used in the workshops of Laecanius and Crispinilla.

Table III shows the rims which are products of these two workshops. One of these (No. 113) is completely identical in form to the amphora published by Gnirs.[63]

As to piece No. 114, it must be pointed out that this amphora fragment was imprecisely reconstructed earlier. The fabric tests and the more recent excavations in Salla have confirmed that another rim belongs to the handles and the upper part of the amphora, thus the

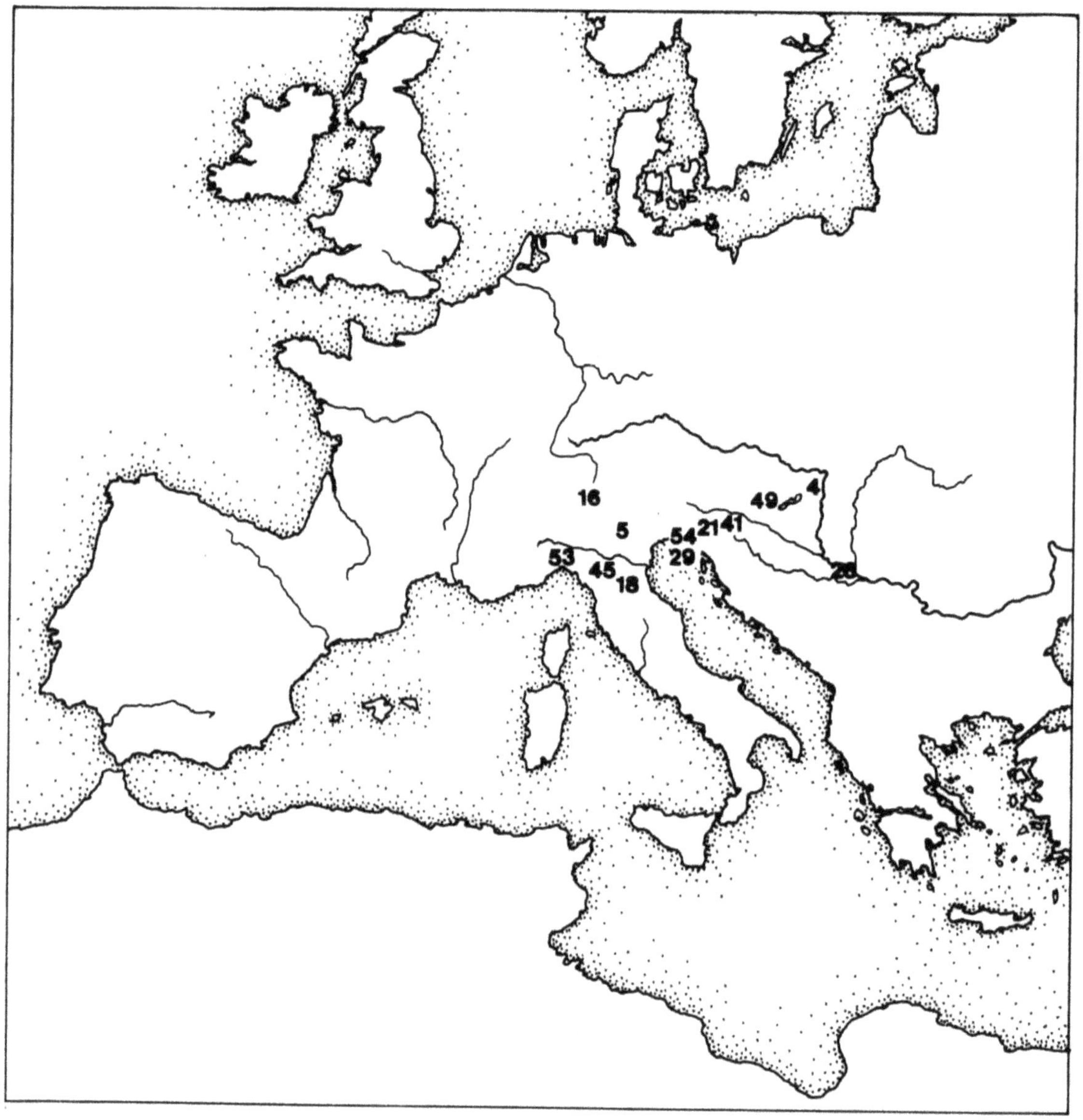

Figure 7 Distribution of the amphorae with imperial stamps

final and correct reconstruction became possible.[64]

The fragments illustrated in Figs. 17-24 and (Nos. 140-148) are also products of these two workshops.

On the neck fragment No. 136, PPC cut perpendicularly may be seen; in addition after firing the number LXI was engraved on the upper part of the body of the amphora Poetovio No. 86. Under the handle of the amphora of Laecanius (No. 52) the number LXXIII may be found.[65] On one of the pieces from Poetovio (No. 83), we find the signs ~~III~~ ~~IIII~~ fairly difficult to perceive.

Characteristic also are the fragments of Poetovio Nos. 91 and 92, along the rim of which a horizontal groove runs. Such small sized amphorae were found in Celeia too (No. 72); this type may also be observed among the finds collected from Loron, housed in the museum at Porec.

III.4. Other Dressel 6B amphorae

The stamps of the following amphorae, which may be classified among the type Dressel 6B, belonged to the sites along the Amber Route (Fig. 8).

CAMI	Nauportus	No. 49
COSAE	Emona	No. 60
APICI	Emona	No. 61
APIC	Emona	No. 62
(VARI PA)CCI	Emona	No. 63
..M	Emona	No. 64
...NAEO	Emona	No. 65
L.INV.PATIE	Emona	No. 66
IVN.PA...	Emona	No. 67
FLAV.FON	Emona	No. 68
L.TREOPTA	Celeia	No. 71
PRIMI	Poetovio	No. 85
APHORVS	Salla	No. 95
A..	Salla	No. 117

The results of the petrological analyses

According to the results of the analyses, the amphorae from Emona, Nos. 66, 67, 62 and 60, contain significant quantities of quartz, of microcline and quartzite. Some plagioclase and a certain amount of carbonate (sparite, micrite) and calcite-based fossils were recognized. The fabric of three amphorae did not contain any zoizite (Nos. 60, 62, 67) but augite was detected in sherds Nos. 60, 66 and 67.

The different pieces Nos. 49, 61, 68, 85) may be classified into a different group. Their properties are the high amount of quartz used for tempering, moderate amount of carbonate, some plagioclase and potash feldspar (sanidin, orthoclas, microcline). The quantity of zoizite and muscovite is significant.

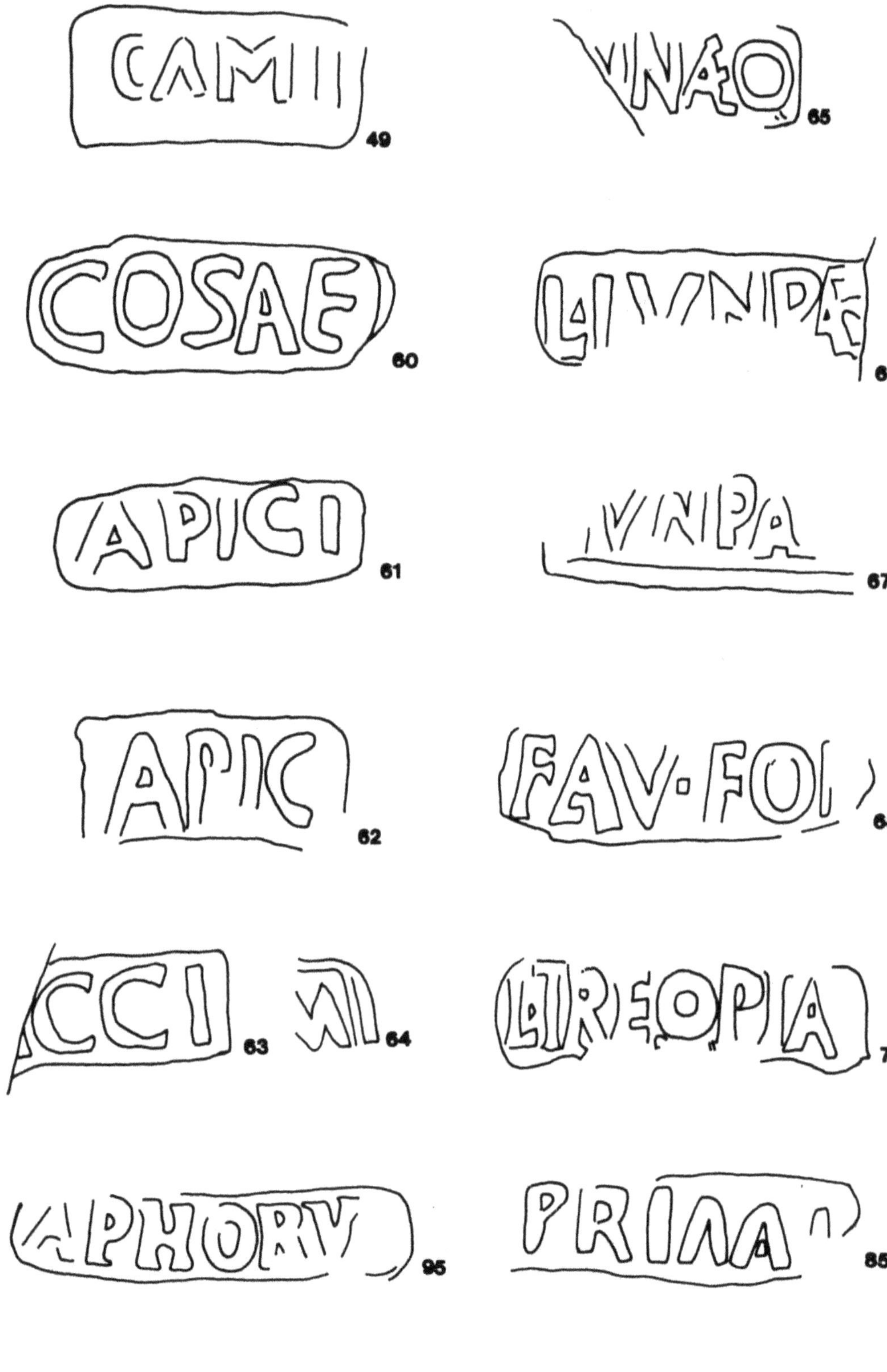

Figure 8 Other Dressel 6B amphorae stamps

The minerals and rocks found in the fabric of the amphorae supplied with the stamps CAMI of Nauportus (No. 49), APICI (No. 61), from Emona, FLAV.FON (No. 68) from Emona, and PRIMI (No. 85) from Poetovio belonging to the latter group, are slightly different from the tempering materials used in the workshops of Laecanius and Crispinilla. It may be supposed that they were made in sites different from the ones above.

The four amphorae, in the fabric of which augite (COSAE (No. 60), L.IVN.PATIE (No. 66, 67) from Emona) and sanidine (APIC No. 62) may be demonstrated were made in workshops where volcanic material was available. So, the places of production of Dressel 6B must have been wider than the Istrian peninsula.

In connection with the inscriptions published in CIL XV.2.,[66] F. Zevi demonstrated that the form Dressel 6, was used for storing oil as well as wine.

Research came closer to the solution when Baldacci and Buchi defined[67] what forms are covered by the Dressel 6 and the "formae 6 similis" types. In connection with the latter type it may also be mentioned that in Aguntum[68] OLEUM HISTR(icum) may be read on one of these forms, while on the Magdalensberg[69] on the same type OLEI HIST(ici) /FLOS/P(ondo) V/Q(inti) /T/VSIDI/I/DEX(tri) may be read. It is connected to this that Gnirs mentions amphorae for storing wine and oil in his publication on the workshop of Fažana. On the basis of the finds excavated so far, we assume that the type Dressel 6B was used for storing oil.

A. Hesnard called the attention of researchers to a Magdalensberg Dressel 2-4 amphora with the stamp of Laecanius[70]. We know for certain that the form was used for wine. This find apparently shows that wine was also transported from this Istrian workshop to the northern provinces.

But the sherd indicated by Hesnard does not exist among the finds of Magdalensberg; probably this assumption was based on the wrong interpretation of the publication. The existence of an Istrian wine trade in the 1st century A.D. needs further evidence, although some of the Dressel 2-4 forms (Nos. 7 and 30) mentioned above, resemble the material of the amphorae produced in Laecanius' workshop.

IV. Dressel 7-11 Nos. 194-215

Spanish garum was transported in vessels of specific forms. These forms are so similar to each other that there is no need to separate them.[71] Their characteristic shape is a rim widening upwards with a long neck. The handles join the oval body, ending in a long spike at the bottom. On some forms long grooves may be found on the upper part of the handle, others have only a simple flat handle.

The colour of the fabric is yellowish orange, in the case of the pieces with coating the colour ranges between a yellowish and a greenish hue.[72]

Among the finds along the Amber Road, on the handle of an amphora from Emona of type Dressel 8, (No. 194) we found the stamp CLANI.AVG+ (Fig. 9; the height of the letters is 11mm).

Stamps are found extremely rarely on these forms, but the same stamp occurs on the Magdalensberg (CLANI.AV...).[73] From the inscription in black paint on one of the pieces from Poetovio (No. 199) we know that the amphora contained garum made in the Spanish way (Pl. 10: G/ari/ F(los) /HISP(anici) /T.THORII.

On another fragment from Poetovio (K 200), a part of an inscription in red paint has been handed down to us: T.TR.A .

On the basis of the inscriptions on this type, Zevi has demonstrated[74] that they were made at the production site, written in black paint, in capital letters, with a careful script similar to that on Dressel 20. They begin with an indication of the product stored in the vessel followed by the ripening date, in such cases there is an indication of the mode of preparation. The inscription ends with the name of the producer in the genitive. Inscriptions in red paint are thought to have been made when the amphorae arrived in the hands of the merchant.

Ancient sources[75] describe the methods of making garum. So do the volumes of CIL compiled by M. Della Corte,[76] Dressel[77] and A. Mau[78]. In an account of the amphorae from Augst, O. Bohn[79] published inscriptions indicating different kinds of garum. Bohn supposes[80], in relation to garum Hispanorum, that Columella, who came from Cadiz (Gades) published an ancient recipe in his work.

The amphorae type Dressel 7-11 is also found in Pannonia and at Vetus Salina[81] and Aquincum.[82]

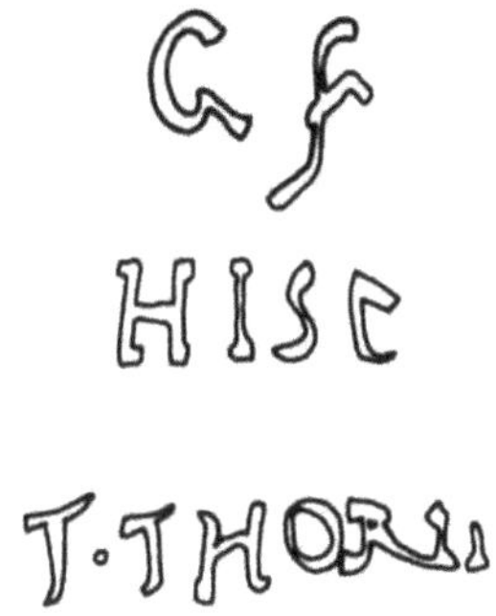

199

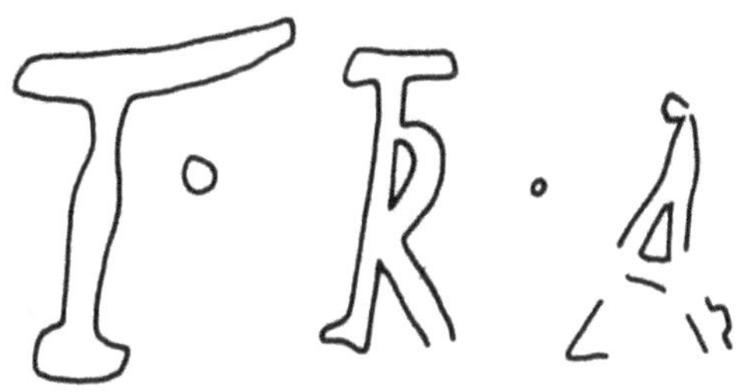

200

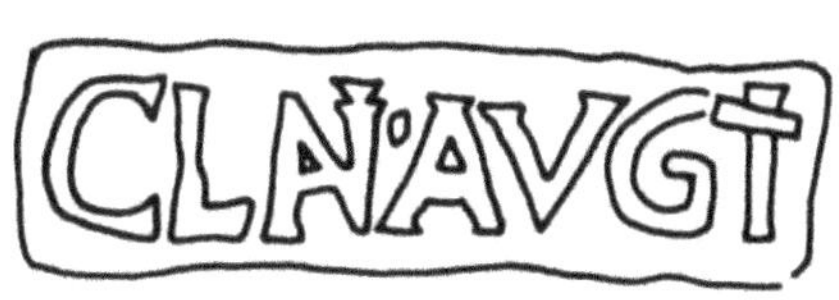

194

Figure 9 Painted inscriptions and stamps on Dressel 7-11 amphorae

V. Dressel 20 (No. 216-223

The oil and olives of Baetica were distributed in characteristic, spheroidal amphorae which were current from Augustus until the middle of the 3rd century A.D.[83] These amphorae, filled with cheap Spanish oil were sent to almost all the provinces of the Empire; thus in Pannonia too a number of pieces of this type are found[84].

The characteristic feature of the form is the spheroidal body, ending in a short spike at the bottom. The profile of the rim, the cross-section of the handles as well as their form, and the line of the neck varied in different periods.[85]

The rim of the oldest types, around the middle of the 1st century is shaped like a ring, the neck is short, the handles are circular and rise higher than the upper joining point of the handle. In Flavian times and the first decades of the 2nd century, the section of the rim is either ovoid or triangular; the neck and the handles are somewhat elongated. From the middle of the 2nd century, the neck becomes shorter again, the handles come closer to the body, and the rim profile becomes quite flat.

Among the finds on the Amber Route are a number of unpublished fragments from Aquileia; one sherd has been published from Emona[86] (No. 216). Abramic published two stamps from Poetovio[87] - although without indicating the type - which are found on handle fragments (Nos. 217, 218). From Salla, in addition to a number of body sherds, we find two rims and a handle fragment. There is only one fragment known from Savaria (No. 223) (Plate 11) which came to light outside the city wall during construction work in the park of the College of Education Szombathely. For lack of picture documentation, it is difficult to precisely define the finds from Carnuntum.[88]

The stamp from Poetovio GMM (No. 218) (Plate 11), is included in a frame <u>tabula ansata</u>, stamped on the upper part of the handle. G.E. Bonsor published such a stamp[89] from El Judio, the district of Arva-Axati, which Callender dated to the second half of the 2nd century A.D.[90] A piece identical with the second stamp from Poetovio (Plate 11) Q.L.S. (No. 217) was found by Bonsor, also in the area of Arva-Axati (Real Tesoro).[91] On the basis of the slightly elongated form of the handle, it may be assigned to the period before the middle of the 2nd century.

The rim fragment from Salla (No. 220) was excavated from the layer 4/A-D; it thus came to the site in the period between the first quarter and the third quarter of the 2nd century. The other fragment from Salla (No. 221) is incomplete; as to its form, it is characteristic of the end of the 2nd century.

The fragment from Savaria - the body, bottom and one handle of which broke off - represents a form characteristic of the end of the 2nd century.

On the basis of these finds, we think that Spanish oil appeared on sites along the Amber Route only in the 2nd century A.D. Even the earliest find cannot be dated before the age of Hadrian i.e. after the decline of the Istrian workshops. Merchants began to supply Spanish oil to compensate[92].

The results of the petrological analyses

With the test we could evaluate three fragments (Nos. 217, 218, 220) according to which the moderate quantities of mica slate and quartzite are characteristic. Muscovite and biotite are present in tiny quantities. Potash feldspar, plagioclase and microquartzite are present in two samples from Poetovio only. Carbonate and calcite fossils appear in moderate quantities. Only in the fabric of two amphorae from Poetovio (Nos. 217 - 218), produced at Arva-Axati, can we detect augite.

VI. Type Pelichet 46 (No. 224-228)

This type[93] is identical with the forms Dressel 38 and Callender 6.

The rim often bends over the handles, the cross-section of the handle is always oval, at the meeting of the neck and the body there is a pronounced carination. The body is shaped like an elongated pear; it ends in a long, spiked bottom, the handles join the neck.

We found an almost whole vessel of this type in Poetovio (No. 224); in Salla we found two smaller fragments (Nos. 225-226) at Savaria, the body and a handle sherd from the same vessel came to light (No. 227).

Among the fragments from Salla, No. 225 was to be found in the layers 5-6 of site "a" and No. 226 in layer 5 of site "t". This means that both sherds were buried in the 1st century A.D.; the specimen from Savaria came from a layer dated to the second half of the 1st century A.D.

In Pannonia this type is found in Gorsium,[94] the camp of Vetus Salina[95] and Aquincum.[96]

On the basis of the formal properties and their material, these amphorae were probably made in Spain.

VII. The Rhodian type (Nos. 229-274)

In this amphora type wine was also transported in the 1st century A.D. Here too the form was derived from Greek predecessors.[97]

The rim protrudes like a sphere, the neck has the shape of a long cylinder. The body protrudes like a ball, then tapers evenly, and ends in a spike at the bottom. The upper part of the handle joins the neck under the rim, then it bends in an acute angle and joins the upper part of the body; its cross section is circular.

Within this type, the following groups could be distinguished in the material from the Amber Route:

a. The form found in Cisalpina, known for its reddish yellow colour and small size (No. 268).

b. This group is characterized by a slim form and thin, rod-like handles (Nos. 238, 240, 242-244, 263-267, 271).

In the case of the following pieces (as they are only handle fragments) the identification was based on the typology of the junction of the upper and lower handle lengths.

c. The upper part of the handle is arched and short (Nos. 234-235).

d. The upper part of the handle is quite short; the lower part starts from a pointed spike (Nos. 237, 257, 258, 270).

e. The upper part of the handle is slightly curved, the lower part starts from a very small spike (Nos. 259-262).

f. A piece of handle showing minimal difference in relation to the previous group. But its fabric is quite distinct (No. 236).

The Rhodian form is found all along the Amber Route.[98] Some of them were made on the Island of Rhodes or in its vicinity.[99] The pieces which imitated them were probably made in Italy.[100]

Only the contexts of the Salla sherds are known; on their basis, it may be supposed they were current in the 1st century A.D.

VIII. Type Schörgendorfer 558 (Nos. 275-294)

The amphorae of this group can be classified on the basis of their formal characteristics as follows:

1. This type was published by A. Schörgendorfer[101] in his catalogue under the number 558.

   The form of the rim profile is tier-like. The neck and bottom are cylindrical; the body consists of two truncated cones standing opposite each other. The handles join the rim horizontally, then bend perpendicularly; they run parallel to the neck, then join the upper part of the body. The cross-section of the handle is flat; it is divided by two ribs on the outside, between them there is a deep groove. On the inner part we find three ribs projecting only slightly. On the upper part of the body, grooves are found in concentric circles (Nos. 275-277, 286-288).

2. This group includes pieces published in northern Italy,[102] there the shape of the handles is more marked than in the previous group; the lower part of the body also has a different shape because it tapers gradually towards the bottom.

3. The shape of the rim profile is not known; the form of the lower part of the body is similar to the first group, but its arch approaches a straight line[103] (Fig. 11; Pl. 13).

4. A small-sized amphora with the properties of type 1 (No. 289, Fig. 27).

5. A small-sized amphora, the profile of the rim cannot be established in this case either. The shape of the body indicates a bottom tapering evenly (No. 285, Fig. 27).

Painted inscriptions

The inscriptions referring to content - with one exception, No. 285 - are in red paint; the letters were drawn carefully. The finds published to date mention green and black olives.

We have one sherd with the inscription OL(iva) /$\overline{\text{ALB}}$(a) /EXD(ulcis) /EXC(ellens) from Celeia (No. 275 Fig. 10), and one with the inscription OL(iva) /$\overline{\text{ALB}}$(a) /EXDUL(cis).. from Poetovio[104] (No. 276 Fig. 10). On one of the finds of Vindobona[105] the inscription, arranged differently, mentions green olives (Fig. 10) ..EXD(ulcis) /..EL(lens) /$\overline{\text{ALB}}$(a).

Amphorae containing black olives are known from several places;

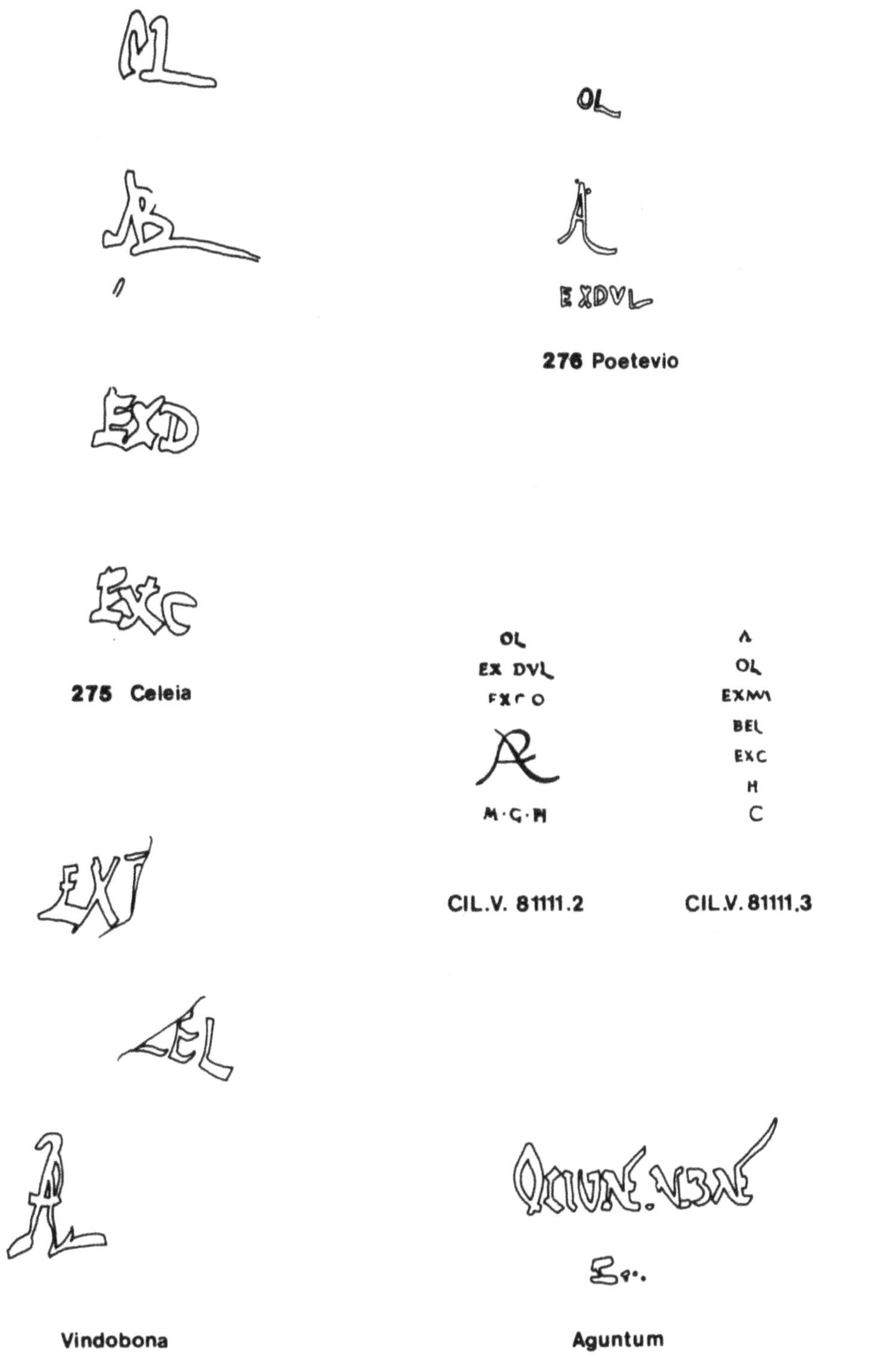

Figure 10 Painted inscriptions - 'green olive' - on Schörgendorfer 558 amphorae

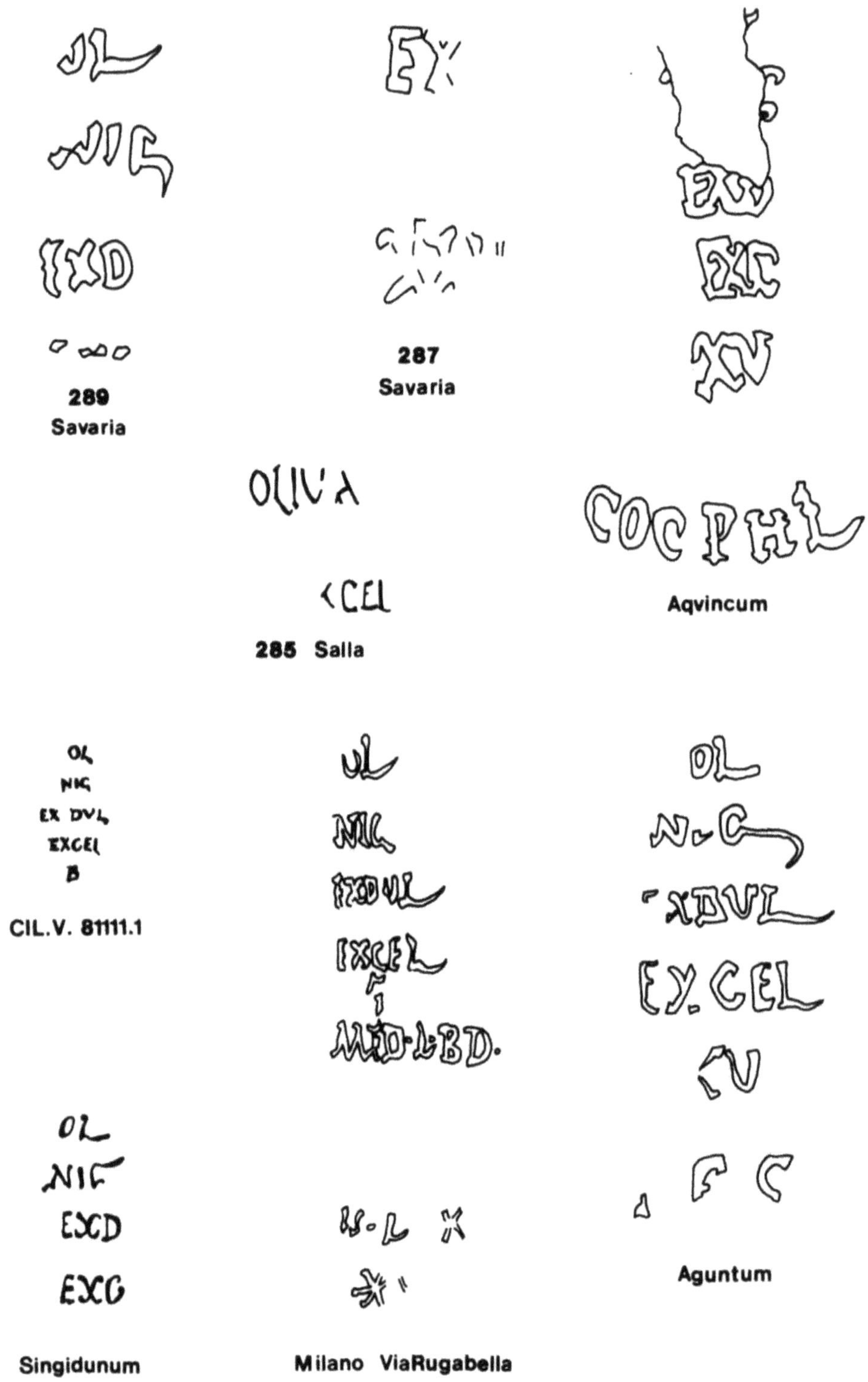

Figure 11 Painted inscriptions - 'black olive' - on Schörgendorfer 558 amphorae

examples with the inscriptions OL(iva) NIG(ra) /EXD(ulcis) /EXC(ellens) were found at Vindobona[106] and Savaria (No. 289 Fig. 11). One fragment (No. 289) came from a small-sized amphora. On one of the Savaria sherds (No. 287), next to the faded letters EX.. on the last line, the number XXII had been written (Fig. 11).

Rather different is the inscription OLIVA/.../EXCEL(lens) written in black paint on a small sherd (No. 285). On the basis of finds from Aguntum, we know that olives were transported in other amphorae as well.[107] (Fig. 10).

Dating

The dating of the finds varies over a wide range of time limits. Schörgendorfer defined the currency of the Vindobona amphora as the 3rd-4th centuries A.D. Baldacci assigned this type to the 1st century A.D. on the basis of a find from Milan.[108]

Among the fragments from Salla, No. 284 came from a ditch in which coins of Claudius minted in A.D. 41 and 43 were found; No. 279 was also excavated from a 1st century context.

Fragment No. 289 from Savaria came to light from a deposit of the 1st-2nd centuries[109]; the remainder of the Savaria amphorae (Nos. 286 and 287) came from excavations of the last century, so we have no data relating to the contexts from which they were retrieved.

We can therefore confirm Baldacci's view that these vessels were in use in the 1st century A.D.

But one find (No. 285) differs from the rest both typologically and chronologically, since it came to light in excavations at Salla from a layer dated between the end of Hadrian's reign and the Marcomanni Wars.

Distribution (Fig. 12)

This type is attested outside the Amber Route at Aquincum[110] and Gorsium[111]; finds are also encountered in Moesia, Noricum and Cisalpina.

The Cisalpina finds published to date were summarized by Baldacci[112]; he says these amphorae are found at Vercelli (CIL V. 8111.2., Fig. 10), Milan (Fig. 10), Chiavenna (CIL V. 8111.1., Fig. 11), Como, Cremona, Asti (CIL V. 8111.3., Fig. 10), Feltre and Aquileia.

In Noricum such amphorae are present at Ovilava[113] (Wels) and Aguntum[114] (Fig. 10), although the form of the Ovilava find is not known. D. Bojović published an unprovenanced sherd from the Singidunum Museum collections[115] (Fig. 15). He adds that several examples of such amphorae were found in the region of Djerdap.[116]

Place of origin

We do not know where these amphorae were made. But in two papers,

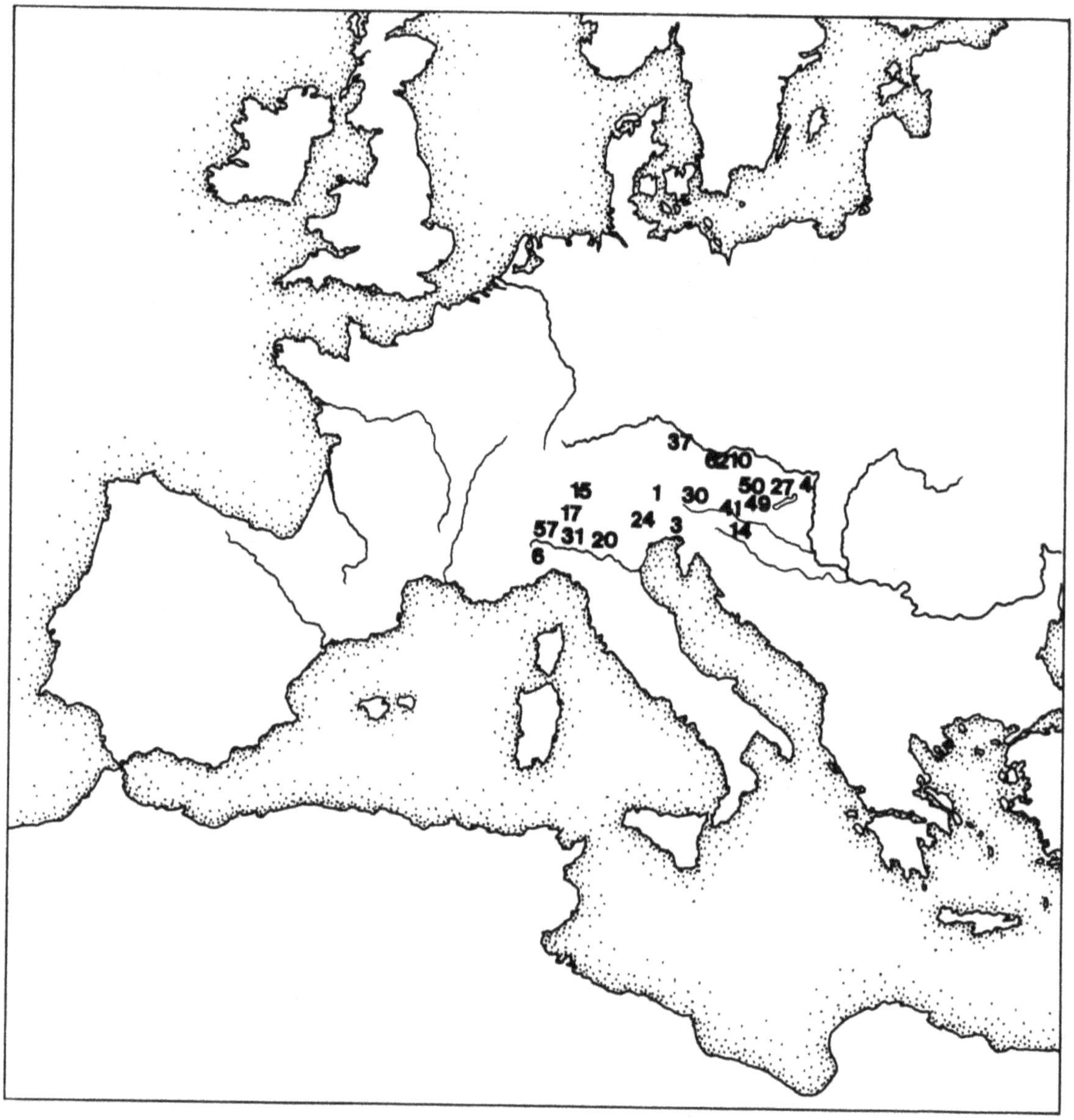

Figure 12 Distribution of the Schörgendorfer 558 amphorae

Degrassi raised the possibility[117] that these amphorae were made in the Istrian Peninsula. Baldacci has accepted this[118].

While collecting material, we did not encounter amphorae of this type among the finds in museums in the Istrian Peninsula at Pula and Porec. Not a single sherd of this type was published by Gnirs from the workshop of Laecanius in Fažana. Regions renowned for olives and oil included not only Istria but Baetica and Venafrum as well[119]. In the valley of the Po River, olive trees grew only in the vicinity of the lakes;[120] in the province of Emilia, including Romagna, oil production was insignificant. Olive trees were grown sporadically in Picenum in the province of Marche, but the latter were so insignificant that they do not even feature in the list compiled by Pliny.

In the globular amphorae from Baetica, olives as well as oil were transported[121]; it is only Venafrum oil exports for which we have no evidence. Therefore most students represent the view that only an assumption may have been over amphasized.[122]

<u>The results of the petrological analysis</u>

On the basis of the analysis of the fabric of the finds, only in some cases may we demonstrate an equivalence between the amphorae type Schörgendorfer 558 and the ones made by the well-known producers of Istria. According to the geological analysis the quantity of the granules used may be said to be significant; it changes between 16.5 - 64%, but the average is 25.1% too (Fig. 24).

The proportion of the material added to the clay in the amphorae of Istria shows the following distribution according to the workshops:

C. Laecanius Bassus	3 - 9.7%
Calvia Crispinilla	2 - 5.5%
P. Iturius Sabinus[123]	6.9%
Pieces with IMP stamps	1 - 8.1%

The analyses of the granules used in the amphorae type Schörgendorfer 558 demonstrated that their significant part was made on volcanic territory, since augite, Ti-augite, egerinaugite, pigenote, as well as volcanic granules were to be demonstrated. These minerals cannot be demonstrated in the fabric of the amphorae of Laecanius, Crispinilla, Iturius and those marked with the IMP stamps.

In the case of two pieces (Nos. 277 and 279) we found a significant quantity of carbonate; these amphorae contain many fossils as well. It is possible that sherds with these components were made somewhere in Istria.

But the carbonate content of the samples got into them after being placed into the early, as a result of saturation, and do not relate to their production. J. Iváncsics, (who did the analyses) indicated central Italy as the origin of those amphorae that have volcanic components. We must undertake further investigations to define the origin of the remainder of the sherds. It is assumed that places about which we do not yet have sufficient information can also be taken into account.

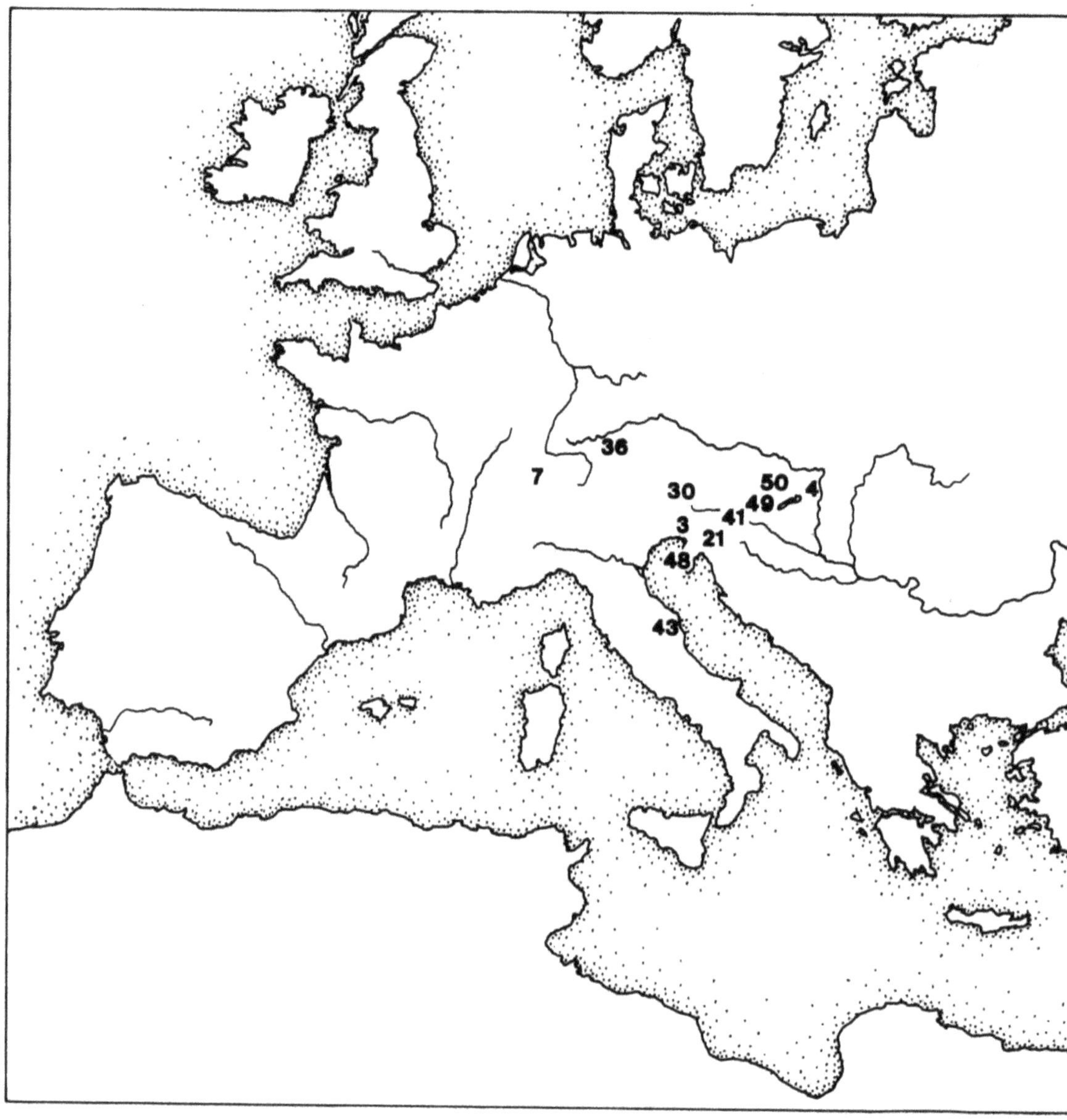

Figure 13 Distribution of the Portorecanatian amphorae

IX. Portorecanation type (Nos. 295-311)

(Amphorae with funnel-like rims and necks)

Among the finds from the Amber Route, a few amphora fragments came to light where the rim and neck take the form of a long funnel. On the basis of the sherds found at Aquileia[124] and Augst[125], the individual form variations could be defined.

In the case of one type, the body is oval and the bottom ends in a button. In the case of the other type, the lower part of the body is elongated and closes straight, joined by a short spiked bottom. In both types the handle, starting from the middle part of the neck, runs a semi-circle and joins the upper part of the body; its cross-section is circular. Outside Aquileia, finds are reported from Emona (Nos. 295-296), Poetovio (No. 297), Salla (Nos. 298-310) and Savaria (No. 311).

In Pannonia we have such amphorae from Aquincum too.[126] The type is regarded as rare. E. Ettlinger published a sherd from Aventicum[127]; a fragment is known from the Magdalensberg too.[128]

L. Mercando excavated several fragments in Portorecanati[129]; a similar form is present at Oberstimm as well[130].

There is also a specimen in the museum of Rovinj,[131] recovered from the sea.

Vessels similar to this slim form are present among the amphorae from Augst.

The distribution is shown in Fig. 13.

Petrological analyses

On the basis of the geological analysis (which include the sherds from Aquileia) it may be affirmed that the quantity of quartz is fairly significant; potash feldspar is always present, in small quantities. The content of biotite and muscovite is significant. That of quartzite is low or absent altogether. Carbonate appears in a significant proportion too - with the exception of No. 302 from Salla, where only traces are present. In one of the Aquileian type (an oval form) we could detect augite too.

The centre of manufacture of this form is not yet known to us; so far only one sherd has been excavated at Portorecanti[132], on the basis of which the workshop could not be located.

On the basis of the preparation and the properties of the fabric, the form could have been made in several workshops.

Chronology

Among the finds at Salla, some (Nos. 301-302, 304) were excavated from layers dating from the middle of the 1st century A.D. to the first quarter of the 2nd century.

Also from Salla, the finds Nos. 298-299. 303, 306 came from a layer dated from the last quarter of the 1st century A.D. to the first quarter of the 2nd century. Mercando dated the finds excavated from the graveyard at Portorecanti to between the reigns of Domitian and Antoninus Pius.[133] But the Magdalensberg sherd can be dated before the middle of the 1st century A.D.

X. The African type (No. 312)

Only one amphora of the so-called African type was found among the finds from the Amber Route. It has a flared rim and short neck; both small handles have flat cross-sections. Below the neck the body becomes wider; it reaches its largest diameter close to the bottom. The bottom ends in a small, pointed spike.

These amphorae probably served for transporting oil from North Africa; their precise centre(s) of manufacture cannot be defined at present.[134] They are thought to have been current in the 5th and 6th centuries A.D.

XI. The Spatheion type (No. 313)

The expression Spatheion is used for a group of amphorae which have a flared rim, a narrow long body and a bottom ending in a pointed spike. The neck is relatively long; there are two short handles. We found a specimen in Poetovio; at Aquileia several more such amphorae were excavated. On the basis of V. Grace's[135] suggestion, we suppose that wine was transported in the form, but J.A. Riley[136] prefers to keep an open mind on the contents.

XII. Type Bónis XXXI/5 (Nos. 314-321)

This group includes small-sized amphorae with a fine finish. They are characterized by an arched rim profile, a narrow neck, and a tubular body. The handles join the body below the rim and at the upper part of the body; their cross-section is circular.

E. Bónis was the first to publish amphorae of this type, from the 1st century A.D. cemetery at Emona (Nos. 314-315)[137]. Two pieces found at Savaria are identical (Nos. 320-321). We know a similar form from Salla (No. 318); it differs from the rest only in its fabric and colour. In the reserve collections of the museum at Aquileia are other speciments of the type.

XIII. The Aegean type (No. 323)

Ostia VI. Niederbieber 77. Kapitän II. Zeest 79.[138] Its distinct typological feature is the horizontal grooving over the whole body. The rim and neck are cone-shaped; the upper part of the neck is flat, the lower part ends in a cone tapering downwards. The thick, flat, grooved handles run from the neck; they are sharply bent at rim height and run vertically down to the upper part of the body. The thick wall of the amphora is red.

Among the finds from the Amber Route, we find handle fragments from this amphora type (No. 323) only at Scrabantia.

These vessels were in use from the middle of the 3rd century A.D.[139]

In Pannonia, several fragments of this type have been reported from Aquincum.

There follows a list of stamps from amphorae whose form is not recorded.

AVLVS	CIL III. 12010.6.	Poetovio
ZOSMS or ZOEN	CIL III. 12010.30.	Poetovio
EVPOL	CIL III. 14371. 6.	Poetovio
LI	CIL III. 12010.44.	Poetovio
TIMO+	CIL III. 14371.6.B.	Carnuntum

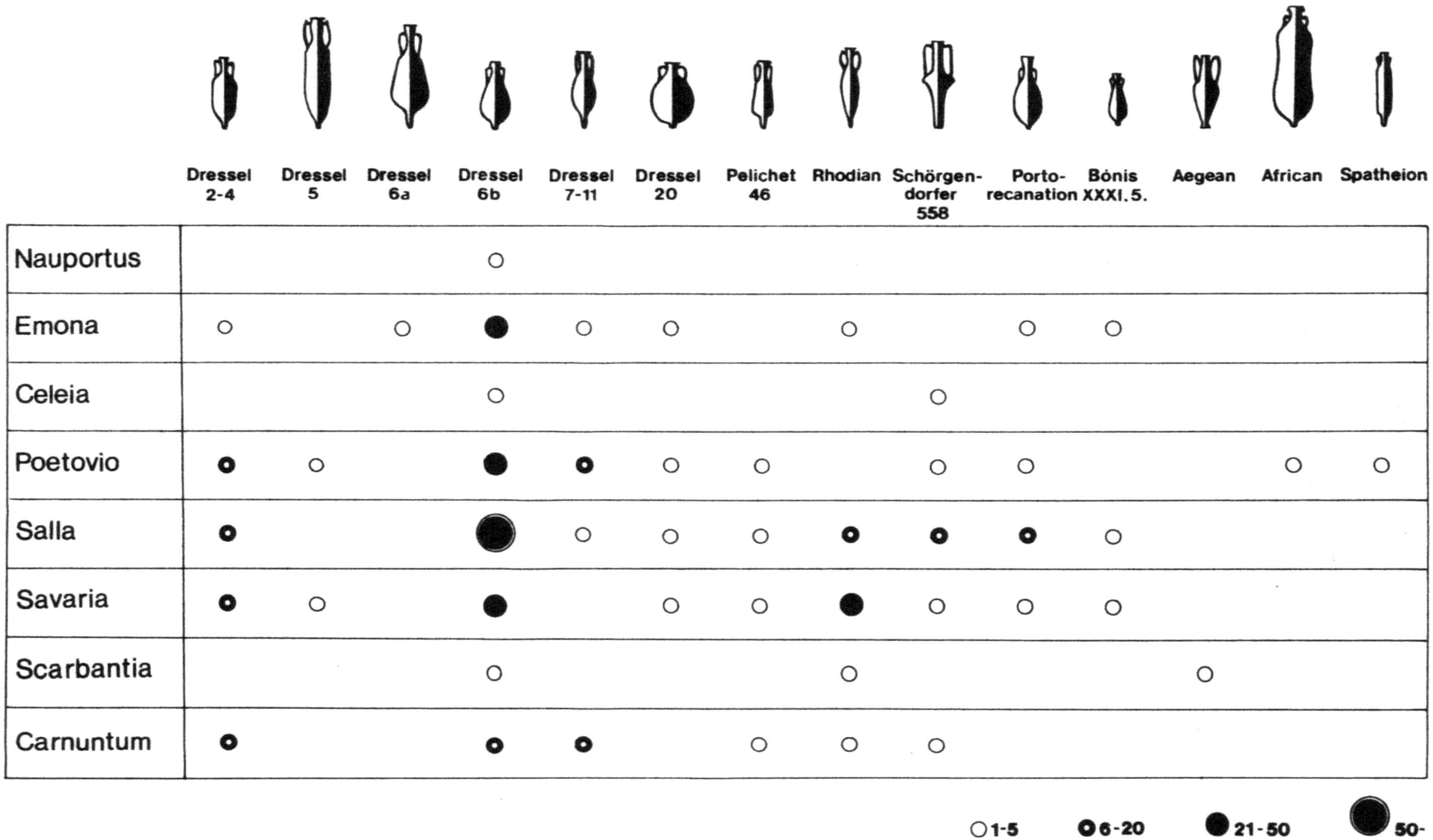

Figure 14 Amphorae types according to sites

SUMMARY

The large number of amphorae dug up along the Amber Route to date, as is apparent from the list, are characteristic finds of the early period of the Roman Empire. Their sites indicate the trade relations with different provinces, and their form, stamps and inscriptions make it possible to identify the place of manufacture as well as to trace the trade in wine and other foodstuffs imported by Pannonia.

The economic changes of the two centuries preceding the occupation of Pannonia exerted their influence here too.

In the 2nd century B.C., after a reduction in the commercial role of Rhodes, following the devastation of Corinth and Carthage, Italian tradesmen conquered the markets of the eastern Mediterranian. This brought about the deterioration of the economic and commercial centres and Asia Minor. The crises caused by the wars against Mithridates were felt in the exchange of goods in the Adriatic and the Aegean regions as well. Through their capital investment, Italians promoted the formation of new centres of production.[140]

At the same time it was fashionable for certain members of the Roman nobility to invest in northern Italy part of the riches they had obtained in the East. At the end of the Republic, Istria also attracted senatorial and equestrian capital. As is apparent from the archaeological evidence reviewed here, the output of the Istrian territories was of major importance in the trade of Pannonia.

The provinces beyond the Alps - including Pannonia - offered good markets; following the conquests firm trade connections were established in their final form. The existing trade routes of antiquity - with a starting point at Aquileia - made it possible to send the goods to their markets.

Amphorae were transported to settlements along the Amber Route. Although Strabo remarked that from the Nauportus Entrepot[141] those products stored in amphorae and barrels were transported further by ships. This should probably be interpreted as transportation to the south-eastern part of the province.

Regular trade may be reckoned to have started under Tiberius, and certainly no later than Claudius. This meant that the inhabitants of Pannonia, more precisely the Italian population resident there, were always able to consume foodstuffs produced in the Mediterranian.

In the amphorae transported here, wine, oil, olives and fish sauce were stored.

Amphorae used for storing wine are indicated in the Dressel table of forms by numbers 2-5 and 43. The use of these types is estimated to have started in the 1st century or by the beginning of the 2nd century A.D. Among the types current from the 3rd century, a form originating from the Aegean region (No. 323), as well as the so-called Spatheion type (No. 313) of the 5th century, were identified.

One of the best-known types of wine amphorae is Dressel 1. It is found throughout the western Mediterranian but on sites along the Amber Route it occurs nowhere other than Aquileia. In Noricum finds have only been published from the Magdalensberg. It is thought that this type was no longer produced at the end of the Republic[142] and it was therefore no longer in use at the time of the conquest of Pannonia.

Among our finds there are several specimens of form Dressel 2-5. This type was made on the Island of Kos, in northern Italy, Campania, Apulia, Gallia Narbonensis, Tarraconensis and, perhaps, in the Istrian Peninsula.[143]

On Dressel 2-5 amphorae with painted inscriptions from the Castra Pretoria at Rome, sherds with the names of consuls can be dated from the beginning of the 1st century A.D., while on the basis of the Ostia finds, an upper limit in the early 2nd century can be postulated.

In all these forms celebrated wines were transported.[144] They included amineum from Campania, baeterrense from southern Gaul, lauronense from Spain, lambrusco, pucinum from northern Italy.

The only ancient source to indicate the significance of the Istrian wine trade is Cassiodorus.[145] Recently it has not been proved, that Dressel 2-4 type amphorae were made in workshop of Laecanius of Fazana.[146] The wines transported in Dressel 6A amphorae may have originated from the eastern coastline of Italy.

It was in Dressel 43, i.e. in the so-called Rhodian type, that average quality wine was exported from Rhodes which (some believe) was consumed mostly by the military. It is assumed that this form was later manufactured in Italy as well. This type is ubiquitous in military and civilian settlements of the 1st century A.D. It is found throughout the Amber Route.

The characteristic feature of the wine amphorae used in the 3rd century are horizontal grooves, the type served for transporting wine from the Aegean.

We have evidence from our finds for the consumption of olives, one of the favourite delicacies of the Mediterranean. It is painted inscriptions on the necks of certain Schörgendorfer 558 amphorae that indicate the storage of first-rate green and black olives in them. A few clear legible inscriptions and the fragments of several amphorae of this type indicate that the military and civil population coming from the Italian or other southern provinces did not have to relinquish the pleasure of their favourite delicacy.

Amphorae for storing oil have been classified by Baldacci, Buchi and Carre. Their classification holds true for finds along the Amber Route as well. For Amphorae of Lamboglia form 2 (the earliest finds are in the museum at Emona) we have no precise information regarding the circumstances of the excavation at Nauportus.[147] Lamboglia 2 is confined to Nauportus and Aquileia along the Amber Route.

Dressel 6B represents a significant number of the finds. In these containers, Istrian olive oil was transported via Aquileia and Nauportus to the Amber Route settlements from the beginning of the 1st century.

When the production of Istrian oil declined after the first quarter of the 2nd century A.D., in the settlements along the Amber Route and throughout Pannonia, the population turned to the consumption of Spanish oil (as is apparent from the incidence of Dressel 20 amphorae).

The fish sauces made in Spain also featured in the diet of the folk who settled down in Pannonia, as proof we find amphorae which contained spiced sacues prepared from different types of fish at Emona, Salla, Savaria and Carnuntum.

The inscription painted in black on one of the amphorae from Poetovio (No. 199) mentions garum made in the Spanish way. An inscription painted in red found on another sherd probably indicates the name of the tradesman (No. 200). We rarely find stamps on amphorae of form Dressel 7-11. There is a sherd from Emona (No. 194), on the handle of which is the stamp of a workshop which has not yet been identified. It is well-known that the fish sauces made in Spain were transported throughout the Empire in the 1st century A.D. No obstacle prevented these products reaching any province, unlike Spanish oil which then had to contend with the serious competition posed by Istrian production, particularly in the provinces beyond the Alps. No wonder therefore that in Pannonia until the early 2nd century A.D., one finds, besides Istrian oil amphorae, only Dressel 7-11 fish sauce amphorae from Spain. The characteristic globular amphorae from Baetica only appear after the decline in Istrian production. In addition to garum, other marine products were also transported to Pannonia. Along the Amber Route, Salla,[148] (in the eastern part of the province) and Aquincum,[149] are sites with finds indicating the consumption of oysters.

Amphora finds along the Amber Route may be dated primarily from the early period of the Roman Empire; a few sherds, however, were current in the late Roman period.

the quantified proportions of the identifiable amphorae express the quantities of products transported therein, as follows:

wine	29%
oil	48%
olives	6%
fish sauce	7%
other products	10%

Understandably, the consumption of oil is the highest of all. Broadly speaking, similar proportions are apparent in the case of the amphorae from the Magdalensberg and Cisalpina as well. Likewise the material from the early Imperial villa of San Potito at Ovindoli in Central Italy.[150]

The amphora finds from Noricum, Pannonia, and Cisalpina reflect similar commercial influences. San Potito however showed characteristic features expressive of the catchment area of the Western Mediterranian. Dressel 6A and 6B are absent there. It may be stated that in the provinces beyond the Alps and in Cisalpina the use of Istrian oil, as at San Potito that of Baetican was standard until the end of Hadrian's reign. After that, however, Spanish oil can be recognised until the middle of the 3rd century A.D.

Wine arrived in both territories from Italy, Spain and the Greek islands.[151]

Fish sauces were transported from Spain. The products transported in amphorae were necessary for everyday cooking and meals. The appropriate trade connections were established for regular deliveries. The traffic through Aquileia was managed by houses of tradesmen with the big producers of Spain, Gallia-Narbonensis, Italy, the Greek islands and later perhaps Africa.

The quantity of finds revealed along the Amber Route indicates that the deliveries could only satisfy the requirements of the Italian population. Only very infrequently have amphorae been published from sites representing the native population.[152] We know from Cassius Dio[153] - although from a later period - that the Pannonians, in the view of the Romans, lived under very meagre circumstances: they were unlucky too because their homeland's soil and climate could not support olives or the vine. With the barley and millet consumed by them they drank something referred to as wine. On that basis, therefore, they lacked all those things that belonged to a decent way of living.

Among the amphorae from settlements along the Amber Route, we have sherds that may be dated from the earliest times at Nauportus and Emona. This reflects the proximity of Italy and is in fact demonstrated by the amphora stamps. Only part of the stamps found at Emona are present at Aquileia, the Magdalensberg and Siscia; they are not found on other sites along the Amber Route. Since these stamps feature on Dressel 6B, one may postulate that they are earlier than amphorae from the workshops of Laecanius or Crispinilla.

From the first quarter of the 1st century A.D., one of the main suppliers of Noricum and Pannonia was Laecanius Bassus, followed by Calvia Crispinilla; in this period oil was not transported from other territories and workshops.

The stamps COSAE, VARI PACCI, and L.IVN.PATIEN are found at Aquileia,[154] Emona,[155] the Magdalensburg[156] and Siscia.[157] The stamp FLAV.FONT. has been identified at Emona[158] and the Magdalensberg.[159] These amphorae did not find their way to other settlements in Pannonia because they were only current in the period before the arrival of the

Romans. Thus it is possible that Dressel 6A amphorae stamped L.TAR.RVFI. may be found in Siscia (CIL III. 12010.30) as well as the sites of Aquileia and the Magdalensberg.

It may be said in conclusion that, as with other provinces of the empire, in Pannonia (or, rather, in its western parts), people from Italy who settled there regularly consumed foodstuffs from the south. They reached settlements in the province by regular commercial traffic in the 1st and until the 2nd century A.D.; later, the scale of imports declined and only sporadic finds are known to us.

CATALOGUE

The following items can be found in museums of:

Emona	-	Mestni Muzej, Ljubljana, Yugoslavia
Celeia	-	Pokrajinski Muzej, Celje, Yugoslavia
Poetovio	-	Pokrajinski Muzej, Ptuj, Yugoslavia
Salla	-	Göcseji Múzeum, Zalaegerszeg, Hungary
Savaria	-	Savaria Múzeum, Szombathely, Hungary
Scarbantia	-	Liszt Ferenc Múzeum, Sopron, Hungary
Carnuntum	-	Museum Carnuntinum, Bad Deutsh-Altenburg, Austria

With the exception of no. 49 which are in the Narodni Muzej in Ljubljana, and no. 22 which is in Savaria Múzeum, Szombathely.

Abbreviations

inv.no.:	-	inventory number
s	-	site
H	-	height
D	-	diameter of the rim
s.h.	-	section of the handle
th.r.	-	thickness of the rim
d	-	diameter
fr	-	fragment
J.P.I.R.	-	Járdányi Paulovics István Romkert (Open air museum)
R.C.	-	Romkert Cukrászda (part of J.P.I.R.)

DRESSEL 2 - 4

Emona

1. Rim, neck, body fr. and two handles, inv.no.: -,
H=29.5cm, D=13.6cm, s.h.=5.9x3cm

2. Neck, body fr. and base, inv.no.: 3271.
cf. Plesničar-Gec 1972. Grave 129. T.XXXIV. no. 1.

3. Rim, neck, body and handle fr., inv.no.: 1462.
cf. Plesničar-Gec 1972. Grave 612. T.CXLI. no. 3.

4. Rim, neck, body fr. inv.no.: G 1247.
cf. Plesničar-Gec 1972. Grave 668. T.CLIV. no. 6.

5. Rim fr. inv.no.: 2242.
cf. Plesničar-Gec 1983. T.35. no. 1.

Poetovio

6. Complete (restored), inv.no.: 12848.
s: Spodnja Hajdina, Parc St. 1070/2. 4.□. from a depth of
30-60cm, H=110cm, D=13.6cm, s.h.=6.5x3.2cm, d=31.3cm
Pl. 1.

7. Complete (restored), inv.no.: -,
pinkish grey (7.5 YR 6/2), white surface, H=84.2cm, D=10.4cm,
s.h.=4.3x2.5cm, d=25.3cm
Pl. 1.

8. Rim, neck, body fr. inv.no.: 3684.
reddish yellow (5 YR 7/6), H=37cm, D=14.2cm
cf. Mikl Curk 1976. 47, 97,
Pl. 2.

9. Rim, neck, body fr. handles, inv.no.: 3683.
light red (2.5 YR 6/8), H=30cm, s.h.=5.4x2.8cm
cf. Mikl Curk 1976. 47, 97,
Pl. 2.

10. Rim, neck, handle fr., inv.no.: 3685.
yellow (10 YR 8/6), H=12.9cm, D=11.8cm, s.h.=5.3x3cm
cf. Mikl Curk 1976. 47, 97,
Pl. 2.

11. Rim, neck fr., inv.no.: 3681.
very pale brown (10 YR 7/4), H=14.1cm, D=12.1cm
cf. Mikl Curk 1976. 47, 97,
Pl. 2.

12. Stamped handle, inv.no.: R 500.
light red (2.5 YR 6/6), stamp: CSENTISEX...AT,
H=13cm, s.h.=2.3cm
cf. Mikl Curk 1976. T. VII.
Pl. 2.

Salla

13. Handle fr. inv.no.: 83.5.55.
cf. Bezeczky 1984. 161. Pl. 10. no. 54.
fig. 15.

14. Handle fr. inv.no.: 83.5.56.
cf. Bezeczky 1984. 162. Pl. 10. no. 56.
fig. 16.

15. Rim, inv.no.: 83.5.88.
s: between the quadrates D/VIII. and E/IX, area "c", layer 4/D - 6, light red (2.5 YR 6/6), H=7.8cm, D=11.5cm, th.r.=1.3cm
fig. 16.

16. Handle fr. inv.no.: 83.5.104.
stray find, red (2.5 YR 5/6), H=6.1cm, s.h.=3.1x4.2cm
fig. 16.

17. Handle fr. inv.no.: 83.5.105.
s: Northern part of quadrate T - C/V, from yellow clay, from a depth of 190 - 210cm, layer 5, area "t", light reddish brown (5 YR 6/4), H=31.3cm, s.h.=4.9x2.9cm
fig. 16.

18. Rim, inv.no.: 83.5.103.
s: from a kiln in the NW part of quadrate F/III, from a depth of 170cm at the fireplace, 60cm below the brick line in the SW part of quadrate G/XI, layer 5, area "h", very pale brown (10 YR 7/4), H=5.3cm, D=14cm, th.r.=1.8cm
fig. 15.

19. Rim, inv.no.: 83.5.135.
s: SW part of quadrate F/III, from a depth of 180-200cm, layer 5-6, area "a", light red (2.5 YR 6/8), H=4.5cm, D=12.4cm, th.r.=1.7cm
fig. 15.

20. Rim, inv.no.: 83.5.106.
s: quadrate T - C/II, south of the square wall, above the gravel layer at a depth of 80-100cm, layers 4-5, area "t", very pale brown (10 YR 7/4), H=5.7cm, D=12cm
fig. 16.

21. Handle fr. inv.no.: 83.5.53.
cf. Bezeczky 1984. 161, Pl. 10. no. 53.
fig. 16.

Ad-Arrabonem

22. Rim, handle fr. inv.no.: 81.2.26.
s: Katafa-Kavicsdomb, E part of the ditch, from a depth of 150cm, very pale brown (10 YR 7/4), H=5cm, D=12cm, s.h.=5.7x2.1cm
cf. Cserményi - Tóth 1979. fig. 15. no. 5.
fig. 16.

Savaria

23. Rim, handle fr. inv.no.: 76.2.72.
s: Kertész street, OTP house, end of the N. foundation, from a depth of 100-140cm in a charcoal pit; red (2.5 YR 5/8), H=9.6cm, D=14.5cm, th.r.=1.8cm
fig. 15.

24. Rim, inv.no.: 71.1.2264, 71.1.2265.
s: Cinema, Kőszegi street, quadrate No. 251, from a depth of 310-330cm in a brown mortar layer; light red (2.5 YR 6/8), H=8.5cm, D=12.5cm, th.r.=1.7cm
fig. 15.

25. Rim, inv.no.: 67.9.83.
s: Rákóczi street, from trench no. 14, beside the synagogue; very pale brown (10 YR 7/3), H=11.7cm, D=12cm, th.r.=1.6cm
fig. 15.

26. Handle fr. inv.no.: 74.4.1370.
s: St. Márton street nos. 32-34, N. part of trench no. 3, at a depth of 120-200cm in brown loose soil intermixed with charcoal; red (10 YR 5/8), H=5.8cm, s.h.=5.2x2.4cm
fig. 15.

27. Handle fr. inv.no.: 77.2.4338.
s: S part of the Town Hall, quadrate no. 9. layer no. 215, at a depth of 310-340cm in debris intermixed with charcoal and fragments of burnt wood daub; light red (2.5 YR 6/8), H=3.4cm, s.h.=4.9x2.7cm
fig. 15.

28. Rim. fr. inv.no.: 61.3.866.
s: J.P.I.R. from the NE part of the building at a depth of 200-240cm; reddish yellow (7.5 YR 7/6), H=4.5cm, D=11cm, th.r.=1.6cm
fig. 15.

29. Rim fr. inv.no.: 66.3.3417.
s: R.C, W part of trench no. 22, between two walls, at a depth of 180-260cm; reddish yellow (5 YR 6/6), H=4.5cm, D=12cm, th.r.=1.6cm
fig.15.

30. Rim fr. inv.no.: 77.2.5941.
s: Town Hall, Bejczi I. street, E side of quadrate no. 13, layer no. 247, at a depth of 280-290cm; reddish yellow (7.5 YR 7/6), H=9.8cm, D=14cm, th.r.=2.2cm
fig. 16.

31. Handle fr. inv.no.: 77.2.1811.
s: Town Hall, quadrate no. 4. layer no. 119, at a depth of 140-160cm, disturbed layer; red (10 R 5/6), H=7.3cm, s.h.=4x2.2cm
fig. 16.

32. Body fr. inv.no.: 76.2.9.
red (2.5 YR 5/8), H=44cm,
cf. Medgyes 1977. T. VIII. no. 2, 180 p.
fig. 15.

Carnuntum

33. Rim, neck, body fr., handles,
cf. Bericht des Vereins Carnuntum in Wien 1903. 70.

34. Rim, neck, handle fr.
cf. Grünewald 1983. T. 44. no. 7.

35. Rim, neck fr.
cf. Grünewald 1983. T. 44. no. 8.

36. Rim, neck, handle fr.
cf. Grünewald 1983. T. 44. no. 9.

37. Rim, neck, handle fr.
cf. Grünewald 1983. T. 44. no. 10.

38. Rim, neck fr.
cf. Grünewald 1983. T. 44. no. 11.

39. Rim, neck fr.
cf. Grünewald 1983. T. 44. no. 12.

40. Rim, neck fr.
cf. Grünewald 1983. T. 44. no. 13.

41. Rim, neck fr.
cf. Grünewald 1983. T. 44. no. 14.

42. Rim, neck fr.
cf. Grünewald 1983. T. 44. no. 15.

43. Handle fr.
cf. Grünewald 1983. T. 45. no. 17.

DRESSEL 5

Poetovio

44. Rim, neck, body fr. handles, inv.no.: -,
brown (7.5 YR 5/4), white surface, on the neck red painted inscription: P..T; H=100cm, D=10.3cm, s.h.=4.5x2.3cm, d=20cm
Pl. 1.

Savaria

45. Neck, handle fr. inv.no.: 69.6.6.
s: Petőfi S. street, at a depth of 400cm, from yellow clay; light red (10 YR 6/8), H=10.8cm, s.h.=5.4x2cm
fig. 16.

DRESSEL 6A

Emona

46. Complete, inv.no.: -,
s: SAZU, outside the city wall, near the port at the river Ljubljanica.
Pl. 3.

47. Rim, inv.no.: F 64-223.
D=15cm.

48. Rim, neck, handle fr. inv.no.: -,
s: Sumi, kv.29.; D=15.6cm.

DRESSEL 6B

Nauportus

49. Rim, neck, handles, inv.no.: -,
s: Bevka Vrhnika; stamp on the rim: CAMI; H=23cm, D=15.3cm, s.h.=3.8x3.4cm
Pl. 3.

Emona

50. Rim, neck, body fr. handles, inv.no.: -,
s: SAZU, outside the city wall, near the port at the river Ljubljanica; light red (2.5 YR 6/8), H=67cm, D=14.1cm, s.h.=4.1x3.6cm, th.r.=1.9cm
Pl. 3.

51. Rim, neck, body fr., base, handles, inv.no.: G 3223.
stamp on the rim: C.LAEk.BAS; SPERATVS; reddish yellow (5 YR 7/6), H=73.2cm, s.h.=4.3x3.6cm
cf. Plesničar-Gec 1972. Grave 116. no. 1. T.XXXII. no. 7.

52. Rim, neck, handles, inv.no.: F 249.
stamp on the rim: IAEk; VIAT; grafitto on the body beside the handle: LXXIII; light red (2.5 YR 6/6), D=14.9cm, th.r.=2cm, s.h.=4.3x3.9cm

53. Rim, neck, handle fr. inv.no.: M 2424.
stamp on the rim: CAK/...; light red (2.5 YR 6/8); H=23.6cm, D=14.8cm
cf. Plesničar-Gec 1983. T. 16. no. 1.

54. Rim, inv.no.: -,
s: Sumi, kv.29; at a depth of 210-230cm; stamp on the rim: CAk; FI\; light red (2.5 YR 6/6), H=5.3cm, D=16cm, th.r.=2.1cm.

55. Rim, inv.no.: F 562.
stamp: AL.CRISPINILLAE; light red (2.5 YR 6/6).

56. Rim, neck, handle fr. inv.no.: -,
s: Jakopičev vrt; stamp on the rim: ....CRISPINILLAE;
light red (2.5 YR 6/6), H=5.4cm, D=14cm, th.r.=2.1cm.

57. Rim, inv.no.: M 841.
stamp: CRI*l*PIN, red (2.5 YR 5/8), H=5.2cm, D=12.8cm,
cf. Plesničar-Gec 1983. T. 7. no. 13.

58. Rim, neck, body fr., handles, inv.no.: -,
s: Sumi; light red (2.5 YR 6/6), D=14.9cm, th.r.=2.3cm, s.h.=3.4x2.9cm.

59. Rim, neck, body fr., handle, inv.no.: -,
s: Drama; reddish yellow (5 YR 6/8), H=26.5cm, D=14.5cm, s.h.=3.7x3.5cm.

60. Rim, inv.no.: F 1932/OJ.
stamp: COSAE; reddish yellow (5 YR 7/8), H=6cm, D=14cm.

61. Rim, inv.no.: 64.208.
stamp: APICI; reddish yellow (5 YR 6/6), D=16.5cm.

62. Rim, neck, base fr. inv.no.: -,
s: Jakopičev vrt; stamp on the rim: APIC; reddish yellow (5 YR 7/6), D=14.5cm.

63. Rim, neck fr. inv.no.: M 2296.
stamp on the rim: .... ..CCI, H=16.4cm, D=16cm
cf. Plesničar-Gec 1983. T. 33. no. 10.

64. Rim, inv.no.: M 2397.
stamp: ...M; reddish yellow (5 YR 7/8), H=6.4cm, D=15.4cm
cf. Plesničar-Gec 1983. T. 16. no. 5.

65. Rim, inv.no.: F 64-209.
stamp: ...INAEO; D=15cm, th.r.=2.3cm.

66. Rim, neck fr. inv.no.: F64-214.
stamp on the rim: L.IVNPAT(I)E.; H=7.9cm, D=16cm, th.r.=2.1cm.

67. Rim, inv.no.: -,
s: Jakopičev vrt. At a depth of 100cm; stamp: .IVNPAT(I).;
H=8cm, th.r.=2.2cm.

68. Rim, inv.no.: 1350.
s: Trg. Revolucije Česta I, S 59/II/2; stamp: FLAV.FON;
reddish yellow (5 YR 7/8), H=8cm, D=15cm, th.r.=2.3cm.

69. Rim, inv.no.: M 1881.
H=4.7cm, D=14.6cm
cf. Plesničar-Gec 1983. T. 7. no. 12.

70. Rim, neck, handles, inv.no.: -,
s: Gregorčičeva Maja; stamp on the rim: IMP.AVG.GER;
light red (5 YR 7/8).

Celeia

71. Rim, neck, handles, inv.no.: 3511.
stamp on the rim: L.TREOPTAT; light red (2.5 YR 6/6), H=28cm,
D=14.4cm, s.h.=4.6x2.9cm.

72. Rim, neck, handle, inv.no.: 2895.
s: Breg; red (2.5 YR 5/8), H=23.9cm, D=14cm, s.h.=3.7x2.6cm.

73. Neck, body fr., base, handle, inv.no.: 8316.
s: Gubceva street; light red (2.5 YR 6/6), H=35cm,
s.h.=3.7x2.4cm.

Poetovio

74. Rim, inv.no.: 3435.
s: Zgornji Breg, parc.no. 290; stamp: LA-B; H=5.4cm,
th.r.=2.2cm
cf. Mikl Curk 1976. 44, 94. T. VII. no. 48.
Pl. 4. fig. 3.

75. Rim, inv.no.: 3671.
stamp: FELCI; H=3.6cm, th.r.=2cm
cf. Abramič 1925. 110; Mikl Curk 1976. 46, 96, T. VIII. no. 6.
Pl. 4. fig. 3.

76. Rim, neck, handle fr., inv.no.: 10094.
stamp on the rim: C.LAEK BAS; CLYMEN; reddish yellow (5 YR 6/6),
H=12.3cm, D=12.8cm, s.h.=3.3x4.6cm
Pl. 4. fig. 3.

77. Rim and neck fr. inv.no.: 3562.
stamp on the rim: CAL.CRISPINILLAE; H=8.5cm, D=16cm, th.r.=2.5cm
cf. Mikl Curk 1976. 45, 96.
Pl. 5. fig. 3.

78. Rim and neck fr. inv.no.: 3548.
stamp on the rim: CAL.CRISPINILLAE; D=14cm, th.r.=2.3cm
cf. Mikl Curk 1976. 45, 96, T. VII. no. 53.
Pl. 5. fig. 5.

79. Rim, inv.no.: 3579.
stamp: ....CRISPINILLAE
cf. Mikl Curk 1976. 46, 96.
Pl. 5. fig. 5.

80. Rim and neck fr. inv.no.: 3546.
s: Staničeva Njiva; stamp on the rim: CRISPINILI;
cf. Abramič 1925. 110; Mikl Curk 1976. 45, 96, T. VII. no. 53.
Pl. 5. fig. 5.

81. Rim, neck and handle fr. inv.no.: 10932.
s: Zgornji Breg, parc 293, □. 41, Ruševinska plast; stamp on the rim: CRISPINILI.
Pl. 5.

82. Rim, neck, handle fr. inv.no.: 10953.
s: Zgornji Breg, parc. st. 293. □ 41. stamp on the rim: CRIƧPIN; D=12.7cm.
Pl. 5. fig. 5.

83. Complete (restored), inv.no.: 7534.
s: Hajdina; stamp on the rim: IMP.AVG.GER; grafitto on the body, below the handle: ~~III~~ ~~IIII~~; H=68.5cm, D=13.4cm, d=28.6cm
cf. Mikl Curk 1976. 50, 99.
Pl. 6. fig. 5.

84. Rim, neck, handle fr. inv.no.: 7611.
s: Mindritska Njiva, parc. st. 143; stamp on the rim: IMP.NERVAE.AVG; D=13.4cm.
cf. Mikl Curk 1976. 50, 99, T. VIII. no. 14.
Pl. 4. fig. 5.

85. Rim, neck, fr. inv.no.: 3549.
stamp on the rim: PRIMI; light red (2.5 YR 6/8).
cf. Abramič 1925. 110; Mikl Curk 1976. 45, 96, T. VII. no. 55.

86. Neck, body, base, handles, inv.no.: 12849.
s: Spodnja Hajdina, parc. st. 1070/2. 4; grafitto on the body: LXI; light red (2.5 YR 6/6), H=93cm, d=39.5cm, s.h.=4x3.7cm.
Pl. 6.

87. Complete (restored), inv.no.: 12851.
s: Spodnja Hajdina, ko. parc. st. 1070/2. □ 4; red (2.5 YR 5/8), H=80cm, D=14.5cm, d=36.3cm.
Pl. 6.

88. Rim, neck, handle fr. inv.no.: 12833.
s: Spodnja Hajdina, parc. st. 1070/2. □ 3; very pale brown (10 YR 8/4), D=13cm.
Pl. 3.

89. Rim, neck, handle fr. inv.no.: 12826.
s: Spodnja Hajdina, parc. st. 1070/2.□3; light red (2.5 YR 6/8), D=13.8cm, s.h.=3.2x3cm.
Pl. 3.

90. Body, base, inv.no.: 12850.
s: Spodnja Hajdina, parc. st. 1070/2. □4; pink (5 YR 7/4), H=70cm, d=39.3cm.
Pl. 3.

91. Rim, neck, body fr., handles, inv.no.: 11592.
s: Vičava ko. Krčevina, parc. 650, Roševiska Plast; light red (2.5 YR 6/6), H=50cm, D=13.6cm, d=29.6cm, s.h.=3.5x3.1cm.
Pl. 3.

92. Rim, neck, body fr., handles, inv.no.: 10906.
s: Zgornji Breg, parc. st. 293, □40; reddish yellow (5 YR 6/6), H=57cm, D=12cm, s.h.=3.2x3.4cm, d=23.8cm.
Pl. 3.

93. Neck fr., handle, inv.no.: 8835.
reddish yellow (5 YR 6/8).

94. Rim, inv.no.: –,
s: Zgornji Breg; stamp: AELI.CRIS,
cf. Mikl 1960. 156. T. II. no. 8.

Salla

95. Rim fr., inv.no.: 84.1.1.
s: SE part of the quadrates T – A/V, at a depth of 190cm in black soil, layer 5, area "t"; stamp: APHORVS; light red (2.5 YR 6/6), H=13.4cm, H=7cm, th.r.=2.2cm
fig. 18. Pl. 8.

96. Rim fr., inv.no.: 83.5.91.
s: S part of the quadrate C/VIII, yellow clay debris from under the mortar level, -60, -100cm; layer 4/C–D, area "c", stamp: LAE; light red (2.5 YR 6/6), H=5cm, D=16cm.
fig. 17. Pl. 7.

97. Rim, inv.no.: 84.1.2.
s: quadrates F – F/XIII – D, from SW drain, at a depth of 100–150cm, layers 3 – 5, area "d"; stamp: FELIX.PE; light red (2.5 YR 6/6), H=7.9cm, D=17cm, th.r.=2.3cm
fig. 17. Pl. 7.

98. Rim, inv.no.: 83.5.77.
s: quadrates S – B – C, unstratified find; red (2.5 YR 5/8), H=6.9cm, D=15cm
fig. 17.

99. Rim fr., inv.no.: 83.5.10.
cf. Bezeczky 1984. 158. Pl. 3. no. 10.
fig. 17.

100. Rim fr. inv.no.: 83.5.86.
s: quadrates T - B/I, from the grey layer under the gravel stones between two stone walls at a depth of 140-160cm, layer 5, area "t"; light red (2.5 YR 6/8); H=7.7cm, D=14cm, th.r.=2.1cm fig. 17.

101. Rim fr. inv.no.: 83.5.4.
cf. Bezeczky 1984. 157. Pl. 3. no. 4.
fig. 17.

102. Rim, neck, handle fr. inv.no.: 83.5.7.
stamp on the rim: (IMP)AVGG(ER).
cf. Bezeczky 1984. 157. Pl. 3. no. 3.
fig. 17. Pl. 8.

103. Rim fr. inv.no.: 83.5.5.
cf. Bezeczky 1984. 157. Pl. 3. no. 5.
fig. 17.

104. Rim fr. inv.no.: 83.5.6.
cf. Bezeczky 1984. 157. Pl. 3. no. 6.
fig. 17.

105. Rim fr. inv.no.: 83.5.8.
cf. Bezeczky 1984. 157. Pl. 3. no. 8.
fig. 17.

106. Rim fr. inv.no.: 83.5.9.
cf. Bezeczky 1984. 157. Pl. 3. no. 9.
fig. 17.

107. Rim fr. inv.no.: 83.5.90.
s: SE part of the quadrates T - B/IV, from the black layer of ash; at a depth of 180-200cm, layer 5. area "t"; stamp: FEL.T$\overline{VR}$; light red (2.5 YR 6/8), H=7.6cm, D=14.2cm, th.r.=1.8cm fig. 17. Pl. 7.

108. Rim fr. inv.no.: 83.5.12.
cf. Bezeczky 1984. 158. Pl. 4. no. 12.
fig. 17.

109. Rim, inv.no.: 83.5.87.
s: NE part of quadrate E/XIII, at the entrance to room no. I. from grey earth, 20-40cm below the bricks, layers 3-4, area "d"; light red (2.5 YR 6/8), H=5cm, D=13cm, th.r.=1.5cm.
fig. 17.

110. Rim fr. inv.no.: 83.5.11.
cf. Bezeczky 1984. 158. Pl. 4. no. 11.
fig. 17.

111. Rim fr. inv.no.: 83.5.1.
cf. Bezeczky 1984. 157. Pl. 3. no. 1.
fig. 17.

112. Rim fr. inv.no.: 83.5.78.
s: quadrants S-B/V. from the SW trench at a depth of 250-300cm; layer 5, area "s"; reddish yellow (5 YR 6/8), H=9.5cm, D=15cm.
fig. 17.

113. Rim, neck fr. inv.no.: 83.5.2.
cf. Bezeczky 1984. 157. Pl. 3. no. 2.
fig. 17.

114. Rim, neck, handles, inv.no.: 83.5.13.
s: from the south side of room no. I in burnt clay debris, at a depth of 70-90cm, layer 5; and room no. I. from greyish black earth at a depth of 110-120cm, area "c", reddish yellow (5 YR 6/8), H=28cm, D=14.5cm, s.h.=3.8x3cm
fig. 18.

115. Rim fr. inv.no.: 83.5.79.
s: from the NE part of quadrates T - C/IV, from the grey layer under the thick yellow layer, at a depth of 190-210cm; light red (2.5 YR 6/6), H=11.2cm, D=15cm, th.r.=2cm.
fig. 18.

116. Rim, neck fr. handles, inv.no.: 83.5.17.
cf. Bezeczky 1984. 158. Pl. 4. no. 17.
fig. 18.

117. Rim, neck, handle fr. inv.no.: 83.5.107.
s: Quadrate G/XVI. from the burnt grey layer that filled the NE - SW channel, at a depth of 120-130cm, layers 3-4/B, area "d"; stamp: A...; reddish yellow (5 YR 6/6), H=12.1cm, D=11.5cm, th.r.=1.6cm.
fig. 18. Pl. 8.

118. Rim fr. inv.no.: 83.5.16.
cf. Bezeczky 1984. 158. Pl. 4. no. 16.
fig. 18. Pl. 7.

119. Rim, neck fr. inv.no.: 83.5.14.
cf. Bezeczky 1984. 158. Pl. 4. no. 14.
fig. 18.

120. Rim fr. inv.no.: 83.5.15.
cf. Bezeczky 1984. 157. Pl. 4. no. 15.
fig. 18.

121. Rim fr. inv.no.: 83.5.3.
cf. Bezeczky 1984. 157. Pl. 1, Pl. 3. no. 3.
fig. 29.

122. Rim fr. inv.no.: 83.5.89.
s: quadrate C/VIII, from grey ash at a depth of 170-190cm, layer 5 - 6, area "c"; H=6cm, D=15.2cm
fig. 29.

123. Rim fr. inv.no.: 83.5.129.
s: NW part of quadrate T - B/I. at a depth of 210-220cm in grey ash above the burnt layer, layer 5, area "t"; pink (5 YR 7/4), H=5.5cm, D=11.8cm
fig. 29.

124. Handle fr. inv.no.: 83.5.57.
cf. Bezeczky 1984. 162. Pl. 10. no. 57.
fig. 19.

125. Handle fr. inv.no.: 83.5.58.
cf. Bezeczky 1984. 162. P)l. 10. no. 58.
fig. 19.

126. Handle fr. inv.no.: 83.5.59.
cf. Bezeczky 1984. 162. Pl. 10. no. 59.
fig. 19.

127. Handle fr. inv.no.: 83.5.60.
cf. Bezeczky 1984. 162. Pl. 11. no. 60.
fig. 19.

128. Handle fr. inv.no.: 83.5.61.
cf. Bezeczky 1984. 162. Pl. 11. no. 61.
fig. 19.

129. Handle fr. inv.no.: 83.5.62.
cf. Bezeczky 1984. 162. Pl. 11. no. 62.
fig. 19.

130. Handle fr. inv.no.: 83.5.63.
cf. Bezeczky 1984. 162. Pl. 11. no. 63.
fig. 19.

131. Handle fr. inv.no.: 83.5.132.
s: from the N part of G/XV. and the E part of F/XV, at the top of the channel, at a depth of 40cm, layers 3-4, area "d"; light red (2.5 YR 6/6), H=11.6cm, s.h.=3.7x3.1cm
fig. 32.

132. Handle fr. inv.no.: 83.5.133.
s: from the SE part of quadrates T - C/III-IV, E part of the stone wall, from black layer, at a depth of 230-240cm, layer 5. area "t"; light red (2.5 YR 6/6), H=9cm.
fig. 32.

133. Handle fr. inv.no.: 83.5.136.
s: N part of room no. II. 30-40cm under the brick line, layer 4, area "c"; light red (2.5 YR 6/6), H=8.5cm, s.h.=3.6x2.8cm
fig. 32.

134. Handle fr. inv.no.: 83.5.148.
s: E/III-IV, under the thick yellow layer, layer 5. area "h"; reddish yellow (5 YR 6/6), H=7.2cm.
fig. 32.

135. Handle fr. inv.no.: 83.5.141.
s: room no. I. from filling, layer 4. area "c"; light red (2.5 YR 6/6), H=7.8cm, s.h.=4.3x4cm.
fig. 32.

136. Neck fr. inv.no.: 83.5.92.
s: room no. I. from filling, layer 4. area "c"; grafitto: PPC.. reddish yellow (5 YR 6/6), H=8cm.
fig. 20.

137. Neck, body, handle fr. inv.no.: 83.5.100.
s: quadrate F - J/XIII from brown earth, 60-80cm under the brick line, layer 3-4, area "d"; light red (2.5 YR 6/8), H=20.2cm.
fig. 20.

138. Neck fr. inv.no.: 83.5.101.
s: room no. I. layer 1/B, ara "c"; reddish yellow (5 YR 7/6), H=12.9cm, d=11cm.
fig. 20.

139. Neck, body fr. inv.no.: 83.5.102.
s: W part of quadrate S - C/I at a depth of 150-170cm, layer 5, area "s"; reddish yellow (5 YR 6/8), H=26.6cm.
fig. 20.

140. Body, base fr. inv.no.: 83.5.111.
s: quadrates S - D/II, burnt ash layer at a depth of 100-120cm, layers 4-5, area "s"; reddish yellow (5 YR 6/6), H=16.4cm.
fig. 21.

141. Body and base fr. inv.no.: 83.5.112.
s: quadrate B/IV at a depth of 20cm, layer 1-2, area "f"; yellowish red (5 YR 7/8), H=12.3cm.
fig. 21.

142. Body and base fr. inv.no.: 83.5.113.
s: quadrate F/XIV from grey soil between two clay levels, 30-60cm from the brick line, layer 3-4, area "d"; light red (2.5 YR 6/8), H=10.6cm
fig. 21.

143. Body and base fr. inv.no.: 83.5.114.
s: stray find; reddish yellow (7.5 YR 7/6), H=7.2cm.
fig. 21.

144. Body and base fr. inv.no.: 83.5.115.
s: SE part of the quadrate E/IV from a pit, layer 4/A, area "h"; reddish yellow (5 YR 6/8), H=7.6cm.
fig. 21.

145. Base fr. inv.no.: 83.5.44.
cf. Bezeczky 1984, 161, Pl. 9. no. 44.
fig. 21.

146. Base fr. inv.no.: 83.5.45.
cf. Bezeczky 1984, 161, Pl. 9. no. 45.
fig. 21.

147. Base, inv.no.: 83.5.47.
cf. Bezeczky 1984, 161, Pl. 9. no. 47.
fig. 21.

148. Base, inv.no.: 83.5.116.
s: NE part of the quadrates T - B/I layer 5, area "t"; light red (2.5 YR 6/8), H=17cm.
fig. 21.

149. Base, inv.no.: 83.5.48.
cf. Bezeczky 1984, 161, Pl. 9. no. 48.
fig. 21.

Savaria

150. Rim, neck, handles fr. inv.no.: 65.52.1.
s: junction of Majakovszkij square and Zrinyi street, at a depth of 150cm; stamp on the rim: ĪAEk; VIAT; light red (2.5 YR 6/8), H=200cm, D=13.6cm.
fig. 17.

151. Rim fr. inv.no.: 64.15.2655.
s: Alkotmány street, Quadrates 87-88 greyish black layer at a depth of 250-280cm; light red (2.5 YR 6/8), H=10cm, D=16cm.
fig. 17.

152. Rim fr. inv.no.: 54.91.13.
s: J.P.I.R.; red (2.5 YR 5/8); H=6.2cm, D=13.5cm, th.r.=2.3cm.
fig. 17.

153. Rim fr. inv.no.: 63.5.389.
s: Sanctuary of Isis, Thököly street, Near the Altar; light red (2.5 YR 6/8), H=6cm, D=15cm, th.r.=2.2cm.
fig. 17.

154. Rim fr. inv.no.: 60.1.139.
s: Rumi street, E wall of the cellar at a depth of 110cm; stamp: ITIMOQꟼMI; .E.SR; light red (2.5 YR 6/8), H=7.8cm, D=14.3fcm, th.r.=2cm.
fig. 17. Pl. 8.

155. Rim fr. inv.no.: 67.9.105.
s: Rákóczi F. street, from trench no. 19 at a depth of 300cm; light red (2.5 YR 6/8), H=7.3cm, D=14cm, th.r.=2cm.
fig. 17.

156. Rim, neck fr. inv.nos.: 65.10.39. and 65.10.131.
s: Rumi street, SW part of quadrate nos. 72-74 beside the W wall at a depth of 90cm, 150cm from the S and W walls of trench no. 3 above the gravel layer at a depth of 50cm; light red (2.5 YR 6/8), H=9.5cm, D=15.5cm.
fig.18.

157. Rim fr. inv.no.: 80.1.2346.1.
s: Sörház street, quadrate no. 83 at a depth of 240-260cm, from debris; reddish yellow (5 YR 7/8), H=5.7cm, D=13cm.
fig. 18.

158. Rim fr. inv.no.: 70.1.1542.
s: Rákóczi street, from mortar layer, at a depth of 120-130cm; reddish yellow (5 YR 6/8), H=5.1cm, D=12.5cm, th.r.=1.9cm.
fig. 18.

159. Rim fr. inv.no.: 54.91.12.
s: J.P.I.R.; light red (2.5 YR 6/8), H=9.7cm, D=12cm, th.r.=2.1cm.
fig. 18.

160. Rim fr. inv.no.: 69.1.1664.
s: Sanctuary of Isis, S part of quadrate no. 1771 at a depth of 90-140cm; red (2.5 YR 5/8), H=5.7cm, D=14cm, th.r.=2.2cm.
fig. 18.

161. Rim fr. inv.no.: 71.1.982.
s: Cinema, Közegi street, quadrate no. 372 at a depth of 240cm, from brownish greyish debris; light red (2.5 YR 6/8), H=6.2cm, D=13cm, th.r.=1.7cm.
fig. 18.

162. Rim fr. inv.no.: 69.1.1370.
s: Sanctuary of Isis, N part of quadrate no. 1769 at a depth of 190cm; light red (2.5 YR 6/8), H=4cm, D=12.5cm, th.r.=1.6cm.
fig. 18.

163. Neck, body, handle fr. inv.no.: 54.167.100.
light red (2.5 YR 6/8), H=28cm, d=27cm.
fig. 20.

164. Neck fr. inv.no.: 65.47.388.
s: Berzsenyi square, quadrats nos. 77-97 from charcoal clay layer at a depth of 330-380cm; light red (2.5 YR 6/8), H=15.4cm.
fig. 20.

165. Neck fr. inv.no.: 69.1.3684.
s: Sanctuary of Isis, Thököly street, quadrat no. 1751 E side of the trench at a depth of 100cm; reddish yellow (5 YR 6/6), H=5.9cm.
fig. 20.

166. Body and handle fr. inv.no.: 77.2.2037.
s: Town Hall, SE part of quadrat no. 4 at a depth of 250-270cm, from charcoal debris with red burnt clay layer; light red (2.5 YR 6/8), H=14.2cm.
fig. 20.

167. Body and handle fr. inv.no.: 64.15.1251.
s: Alkotmány street, quadrates nos. 43-63 at a depth of 180-250cm, brick and stone filling; red (2.5 YR 5/8), H=8.9cm.
fig. 20.

168. Handle fr. inv.no.: 66.3.3919.
s: J.P.I.R. trench no. 21 at a depth of 250cm, light red (2.5 YR 6/8), H=9cm, s.h.=3.1x3cm.
fig. 19.

169. Handle fr. inv.no.: 66.3.3470.
s: R.C. trench no. 21 from the debris at the N part of the building at a depth of 80-150cm; light red (2.5 YR 6/8), H=9.6cm, s.h.=3.4x3cm.
fig. 19.

170. Handle fr. inv.no.: 64.15.436.
s: Alkotmány street, quadrate no. 85 inside the building, from wall debris at a depth of 120-160cm, red (2.5 YR 5/8), H=8.2cm, s.h.=3x2.6cm.
fig. 19.

171. Handle fr. inv.no.: 69.6.10.
s: Petőfi street, middle of the building, from yellow clay at a depth of 400cm; light red (2.5 YR 6/8), H=13cm, s.h.=3.9x3cm.
fig. 19.

172. Handle fr. inv.no.: 54.19.27.
s: J.P.I.R.; light red (2.5 YR 6/8), H=9.7cm, s.h.=3.4x3cm.
fig. 19.

173. Handle fr. inv.no.: 62.2.406.
s: Thököly street, quadrate no. 411 at a depth of 200-230cm; light red (2.5 YR 6/8), H=6.5cm, s.h.=3.7x3.1cm.
fig. 19.

174. Handle fr. inv.no.: 59.1.226.
s: Sanctuary of Isis, quadrate no. 369 from under gravel layer no. 2 near the wall; Red (2.5 YR 5/8), H=5.5cm.
fig. 19.

175. Neck and handle fr. inv.no.: 62.9.928.
s: Thököly street, quadrate no. 364 at a depth of 60-120cm; reddish yellow (5 YR 7/8), H=7cm.
fig. 19.

176. Neck and handle fr. inv.no.: 76.2.73.
s: Kertész street, OTP house, from the N trench at a depth of 100-140cm; very pale brown (10 YR 8/3), H=5.4cm, s.h.=5.1x2.7cm.
fig. 19.

177. Neck fr. inv.no.: 66.3.2544.
s: R.C., S part of trench no. 19, from the building at a depth of 280-300cm; reddish yellow (7.5 YR 7/6), H=6.9cm.
fig. 19.

178. Handle fr. inv.no.: 54.91.32.
s: J.P.I.R.; light red (2.5 YR 6/8), H=18.5cm, s.h.=4.2x3.2cm.
fig. 19.

179. Handle fr. inv.no.: 54.91.31.
s: J.P.I.R.; light red (2.5 YR 6/8), H=15cm, s.h.=3.7x3.2cm.
fig. 19.

180. Handle fr. inv.no.: 64.15.389.
s: Alkotmány street, quadrate no. 65 outside the N wall at a depth of 180-220cm, above the road; reddish yellow (5 YR 6/6), H=6.8cm, s.h.=3.6x2.9cm.
fig. 19.

181. Handle fr. inv.no.: 71.1.1403.
s: Cinema, Kőszegi street, W part of quadrate no. 432 at a depth of 400cm, from clay soil; reddish yellow (5 YR 6/6), H=17.8cm, s.h.=3.9x3.3xm.
fig. 33.

182. Rim fr. inv.no.: -,
stamp: AELĪ.CRĪS; light red (10 YR 6/8), H=5.2cm, D=12.6cm.
fig. 5. fig. 29.

Scarbantia

183. Rim fr. inv.no.: Sc. 3a.
s: Town Hall, from a wall in a grey layer at a depth of 150-200cm; light red (2.5 YR 6/8), H=5.2cm, D=14cm, th.r.=2cm.
fig. 29.

184. Handle fr. inv.no.: 55.111.28.
s: Town Hall; yellowish red (5 YR 5/8), H=21cm, s.h.=3.7x3.5cm.
fig. 32.

Carnuntum

185. Rim, neck fr.
cf. Grünewald 1983. T. 45. no. 6.

186. Rim,
cf. Grünewald 1983. T. 45. no. 7.

187. Rim, neck fr.
cf. Grünewald 1983. T. 45. no. 8.

188. Rim, neck fr.
cf. Grünewald 1983. T. 45. no. 9.

189. Rim, neck fr.
cf. Grünewald 1983. T. 45. no. 10.

190. Rim, neck fr.
stamp on the rim: CAL.CRISPINILLAE
cf. Grünewald 1983. T. 45. no. 11.

191. Rim,
cf. Grünewald 1983. T. 45. no. 12.

192. Rim, neck fr.
cf. Grünewald 1983. T. 45. no. 13.

193. Rim, neck,
cf. Grünewald 1983. T. 45. no. 14.

DRESSEL 7 - 11

Emona

194. Complete (restored), inv.no.: -,
s: SAZU, outside the city wall near the port at the river Ljubljanica; stamp on the upper part of the handle: CLANI.AVGTI; very pale brown (10 YR 8/3), H=90cm, D=16.5cm, d=30cm, s.h.=5.5x3cm.
Pl. 9. fig. 9.

195. Rim, neck and handles fr. inv.no.: M 2896.
s: 23, sector C;
pale yellow (2.5 YR 7/4), D=16.5cm, H=27cm.

Poetovio

196. Complete, inv.no.: -,
very pale brown (10 YR 8/3), D=14.9cm, d=30.2cm, s.h.=4.5x2.7cm.
Pl. 10.

197. Rim, neck, body, handle, inv.no.: -,
pale yellow (2.5 Y 8/4), H=72cm, D=18.8cm.
Pl. 10.

198. Rim, neck, body, base fr. handles, inv.no.: 12847.
s: Spodnja Hajdina, Parc. st. 1070/2. □ 4; reddish yellow (5 YR 7/6), H=75cm, D=16.1cm, d=26.8cm, s.h.=4.5x2.9cm.
Pl. 10.

199. Complete, inv.no.: 11385.
s: Staubo Blagounile, Parc. 1148, 1143/1. 1144; yellow (10 YR 7/6), black painted inscription on the neck: G(ari) F(los)/HISP(anici)/T.THORII; H=91cm, D=17.5cm, d=27.8cm, s.h.=5.1x2.6cm.
Pl. 10. fig. 9.

200. Rim, neck, body fr. handles, inv.no.: –,
pink (5 YR 8/4), red painted inscription on the neck: T.$\overline{TR}$.A;
H=32cm, D=17cm, s.h.=5.1x3cm.
Pl. 9. fig. 9.

201. Rim, neck fr. inv.no.: 3682.
very pale brown (10 YR 8/3), H=10.3cm, D=17.2cm.
Pl. 9.

202. Rim, neck, inv.no.: 3535.
H=28cm.
cf. Mikl Curk 1976, 45, 96.
Pl. 9.

203. Body, base fr. inv.no.: –,
pale yellow (2.5 YR 8.4), H=60cm, d=31.5cm.
Pl. 10.

204. Neck, body fr. inv.no.: 12827.
s: Spodnja Hajdina, Zadruga parc. 1070/2. □ 3 at a depth of 50–80cm; very pale brown (10 YR 7/4), H=54cm.

205. Rim, neck, body, handle fr. inv.no.: –,
H=38cm, D=15.5cm, s.h.=5.2x3.1cm, grafitto on the neck: KAΓΠΙ
Pl. 9.

206. Rim, inv.no.: 3689.
reddish yellow (7.5 YR 7/6), H=5.2cm.

Salla

207. Handle fr. inv.no.: 83.5.54.
cf. Bezeczky 1984, 161, Pl. 10. no. 55.
fig. 31.

Carnuntum

208. Rim,
yellow, D=16cm,
cf. Grünewald 1983, T. 44. no. 23.

209. Rim,
cf. Grünewald 1983, T. 44. no. 24.

210. Rim,
cf. Grünewald 1983, T. 44. no. 25.

211. Rim,
cf. Grünewald 1983, T. 45. no. 1.

212. Rim,
cf. Grünewald 1983, T. 45. no. 2.

213. Rim,
cf. Grünewald 1983, T. 45. no. 3.

214. Rim,
cf. Grünewald 1983, T. 45. no. 4.

215. Rim,
cf. Grünewald 1983, T. 45. no. 5.

DRESSEL 20

Emona

216. Rim, body fr. and base, inv.no.: G 4246.
cf. Plesničar-Gec 1972, Grave 969 no. 1.

217. Handle, inv.no.: 3531.
s: Zgornji Breg; stamp: Q.L.S.
very pale brown (10 YR 7/4) H=21cm
cf. Abramič 1925, 110 and Mikl Curk 1976, 45, 96, T. VII.
no. 49.
Pl. 11.

218. Neck fr., handle, inv.no.: 3532.
stamp on the handle: G.M.M.; brown (5 YR 6/3),
cf. Abramič 1925, 110. and Mikl Curk 1976, 45, 96, T. VII.
No. 50.
Pl. 11.

219. Neck, body and handle fr. inv.no.: 3709.
s.h.=4.5x4.3cm,
cf. Mikl Curk 1976, 47, 97
Pl. 11.

Salla

220. Rim, inv.no.: 83.5.80.
s: Quadrats S - E/II. - F/II from debris layer at a depth of 80-120cm, layers 3-4, area "k"; light brown (7.5 YR 6/4), H=5.9cm, D=16cm, th.r.=3.5cm.
fig. 22.

221. Rim, inv.no.: 83.5.93.
s: area "f", stray find; pink (5 YR 7/4), H=7.5cm, D=15cm, th.r.=3.7cm.
fig. 22.

222. Handle fr. inv.no.: 83.5.117.
s: quadrates T - C/IV from black debris soil at a depth of 130-150cm, area "t", brown (10 YR 5/3), H=8.2cm, s.h.=3.8cm.
fig. 22.

Savaria

223. Rim, neck, body fr., handle.
s: from the park of the Teachers' Training College, Szabadság square; pink (5 YR 7/4), D=16.3cm, s.h.=4.8x4cm (information from a private collector).
Pl. 11.

Pélichet 46

Poetovio

224. Complete, one handle broken, inv.no.: 157.
very pale brown (10 YR 7/4), H=86cm, D=16.5cm, d=26.2cm, s.h.=5x2.2cm.
Pl. 9.

Salla

225. Handle fr. inv.no.: 83.5.119.
s: NW part of quadrate -E/II from gravel layer under the kiln at a depth of 220-240cm, layers 5-6, area "a"; very pale brown (10 YR 7/4), H=14cm, s.h.=5.1x2.6cm.
fig. 22.

226. Neck and body fr. inv.no.: 83.5.118.
s: from the S part of quadrate A/IV from gravel layer at a depth of 100-140cm, layer 5, area "t"; very pale brown (10 YR 7/4), H=18cm.
fig. 22.

Savaria

227. Neck, body, handle fr. inv.no.: 77.2.2154; 77.2.2865; 77.2.3789.
s: Town Hall, SE part of quadrate no. 4 from burnt soil, at a depth of 300-320cm; Middle part of quadrate no. 5 from brown charcoal with clay layer at a depth of 320-340cm; W side of quadrat no. 8, from a pit filled with debris at a depth of 330-370cm; Reddish yellow (5 YR 7/6), with pale brown surface (10 YR 6/8), H=51cm.
fig. 22.

Carnuntum

228. Rim, neck, body, base fr. inv.no.: -,
cf. Kubitschek - Frankfurter 1923, fig. 85.
Pl. 9.

Rhodian amphorae

Emona

229. Rim, neck, handles, inv.no.: F 250.
light red (2.5 YR 6/8), H=38cm, D=11.8cm, s.h.=3.4x3.2cm.

230. Rim, neck, handle fr. inv.no.: F 251.
light red (2.5 YR 6/6), H=27.7cm, D=12.3cm.

Salla

231. Rim, neck, body fr. inv.no.: 83.5.29.
cf. Bezeczky 1984, 159. Pl. 6. no. 29.
fig. 23.

232. Rim fr. inv.no.: 83.5.108.
s: from the S part of room I, from the burnt layer at a depth of 70-90cm, layer 4/C, area "c"; reddish yellow (5 YR 6/6), H=4cm, D=12.8cm, th.r.=1.6cm.
fig. 24.

233. Body, base fr. inv.no.: 83.5.37.
cf. Bezeczky 1984, 161, Pl. 8. no. 37.
fig. 24.

234. Handle fr. inv.no.: 83.5.49.
cf. Bezeczky 1984, 161, Pl. 9. no. 49.
fig. 25.

235. Handle fr. inv.no.: 83.5.50.
cf. Bezeczky 1984, 161, Pl. 9. no. 50.
fig. 25.

236. Handle fr. inv.no.: 83.5.51.
cf. Bezeczky 1984, 161, Pl. 9. no. 51.
fig. 25.

237. Handle fr. inv.no.: 83.5.52.
cf. Bezeczky 1984, 161, Pl. 9. no. 52.
fig. 25.

238. Handle fr. inv.no.: 83.5.109.
s: from the NE part of quadrate F/XIV at a depth of 30-60cm, from the grey earth between two layers of clay, layers 3-4, area "d"; red (2.5 YR 5/8), H=10.1cm, s.h.=2.8x2.4cm.
fig. 25.

239. Base, inv.no.: 83.5.110.
s: from the grey ash layer of quadrate S - E/I at a depth of 130-140cm; reddish yellow (5 YR 7/8), H=10cm, d=4.7cm.
fig. 34.

Savaria

240. Neck fr., body, base, inv.no.: 54.321.30.
s: "Gyöngyösi street, Mennyei house" nowadays Köztársaság square; red (2.5 YR 5/8), H=74cm, d=18.6cm.
fig. 24.

241. Rim, neck fr. inv.no.: 75.1.116.
s: J.P.I.R. quadrate no. 91, from under a wall; red (10 YR 5/8), H=17.3cm, D=12cm.
fig. 29.

242. Neck, body fr. inv.no.: 59.1.253.
s: Sanctuary of Isis, quadrate no. 91 at a depth of 320-350cm, under the foundation; red (2.5 YR 5/8), H=13cm.
fig. 24.

243. Neck fr. inv.no.: 64.15.2288.
s: Alkotmány street, quadrate no. 84 at a depth of 50-140cm, from brick and mortar filling; red (2.5 YR 5/8), H=13.8cm.
fig. 24.

244. Base, inv.no.: 64.15.1761.
s: Alkotmány street, quadrats nos. 43-63. from brick mortar filling at a depth of 50-120cm; light red (10 YR 6/8), H=9.5cm.
fig. 24.

245. Base, inv.no.: -,
very pale brown (10 YR 8/4), H=38cm.
fig. 24.

246. Base, inv.nos.: 67.1.66; 67.1.77.
s: The courtyard of the Bishop's Palace, quadrate no. 3 at a depth of 400cm, E part of the NS wall below floor level; light red (2.5 YR 6/6), H=28cm.
fig. 34.

247. Base, inv.no.: 77.2.1611.
s: Town Hall, quadrate no. 3 at a depth of 90-120cm, layer no. 108; light red (2.5 YR 6/8), H=8.1cm, d=4.7cm.
fig. 34.

248. Base, inv.no.: 57.1.854.
s: Thököly street, SE corner of quadrate no. 20 at a depth of 300cm below the layer of debris on the edge of the black layer; red (2.5 YR 5/8), H=19cm, d=5.4cm.
fig. 34.

249. Base, inv.no.: 62.2.1219.
s: Thököly street, E part of quadrate no. 225, W part of quadrat no. 275 at a depth of 100-190cm, mixed layer; light red (2.5 YR 6/8), H=7.8cm.
fig. 34.

250. Base, inv.no.: 54.91.40.
s: J.P.I.R., red (10 R 5/6), H=10.2cm, d=6.3cm.
fig. 34.

251. Base, inv.no.: 74.2.932.
s: Sanctuary of Isis, quadrate no. 2750 at a depth of 200cm, Roman filling; light red (2.5 YR 6/8), H=18.3cm, d=4.2cm.
fig. 34.

252. Base, inv.no.: 59.1.1064.
s: Thököly street, quadrates nos. 211-213, from Roman wall remains at a depth of 160cm; red (2.5 YR 5/8), H=17cm.
fig. 34.

253. Base, inv.no.: 54.91.39.
s: J.P.I.R.; red (2.5 YR 5/8), H=12.1cm, d=5.1cm.
fig. 34.

254. Base, inv.no.: 56.2.253.
s: Sanctuary of Isis, trench no. 1 from grave no. 3 at a depth of 160cm; red (2.5 YR 5/8), H=11.5cm, d=5.5cm.
fig. 34.

255. Base, inv.no.: 54.91.41.
s: J.P.I.R.; reddish yellow (5 YR 6/6), H=10cm, d=6.6cm.
fig. 34.

256. Base, inv.nos.: 65.49.69; 65.59.89.
s: Köztársaság square, foundation trench no. 1, pit no. 2; pale olive (5 Y 6/3), H=34cm, d=2.3cm.
fig. 23.

257. Handle fr. inv.no.: 66.3.3651.
s: R.C. W side of road ditch no. 17 at a depth of 140-180cm; red (10 YR 5/8), H=9.5cm, s.h.=3.2x3.1cm.
fig. 25.

258. Handle fr. inv.no.: 62.9.135.
s: Thököly street, E part of quadrate no. 364 at a depth of 120-160cm, from a debris layer; reddish yellow (5 YR 6/6), H=6.4cm, s.h.=3.3x3.4cm.
fig. 25.

259. Handle fr. inv.no.: 54.91.22.
s: J.P.I.R.; light red (2.5 YR 6/8), H=7.1cm, s.h.=3.5x3.7cm.
fig. 25.

260. Handle fr. inv.no.: 77.2.3239.
s: Town Hall, middle of quadrate no. 7 at a depth of 230-270cm, layer no. 174. disturbed recently; reddish yellow (5 YR 6/8), H=7.6cm, s.h.=3.1x3.7cm.
fig. 25.

261. Handle fr. inv.no.: 54.91.23.
s: J.P.I.R.; reddish yellow (5 YR 6/6), H=11cm, s.h.=3.3x3.5cm.
fig. 25.

262. Handle fr. inv.no.: 64.15.2455.
s: Alkotmány street, quadrate no. 83 at a depth of 120-160cm, from brick filling; light red (2.5 YR 6/8), H=9cm, s.h.=3x3.5cm.
fig. 25.

263. Handle fr. inv.no.: 80.1.2347.
s: Sörház street, quadrate no. 83 at a depth of 160-240cm, from the brown layer filled with debris and charcoal; light red (2.5 YR 6/8), H=20.5cm, s.h.=3.2x3.5cm.
fig. 25.

264. Handle fr. inv.no.: 64.15.2226.
s: Alkotmány street, quadrate no. 88 at a depth of 280-305cm, from black clay soil; light red (2.5 YR 6/6), H=6.2cm, s.h.=3.1cm.
fig. 25.

265. Handle fr. inv.no.: 59.1.1545.
s: Thököly street, quadrate no. 189 from below debris and above the layer of burnt wattle and daub; red (2.5 YR 5/8), H=9.1cm, s.h.=3.2cm.
fig. 25.

266. Handle fr. inv.no.: 59.1.722.
s: Thököly street, quadrate no. 413 at a depth of 160-180cm, below the second gravel layer above burnt layer; red (2.5 YR 5/8), H=11.5cm, s.h.=3.5x3.7cm.
fig. 25.

267. Handle fr. inv.no.: 66.1.1364.
s: R.C. from the N part of "A" trench at a depth of 290cm; red (2.5 YR 5/8), H=18.2cm, s.h.=3cm.
fig. 25.

Scarbantia

268. Complete (restored), inv.no.: 59.3.47.
reddish yellow (7.5 YR 7/6), H=64.2cm, D=6.5cm, d=23.5cm, s.h.=3.2x2.8cm.
fig. 23.

269. Rim, neck, handle fr. inv.no.: 55.111.26.
s: Town Hall; red (2.5 YR 5/6), H=5.7cm, D=13.2cm, s.h.=3.7cm.
fig. 23.

270. Handle fr. inv.no.: Sc. 8.
s: Városháza street, Cinema, from a layer above the Roman road; light red (2.5 YR 6/6), H=8cm, s.h.=2.9cm.
fig. 25.

271. Handle fr. inv.no.: Sc. 6.
s: Városháza street, Cinema; light red (10 YR 5/8), H=7cm, s.h.=3.4x3.3cm.
fig. 25.

272. Base fr. inv.no.: Sc. 4.
s: Városháza street, Cinema, from the Roman house above the roman road at a depth of 50-60cm; red (2.5 YR 5/8), H=9cm, d=5.4cm.
fig. 34.

Carnuntum

273. Handle fr.
cf. Grünewald 1983, T. 45. no. 15.

274. Handle fr.
cf. Grünewald 1983, T. 45. no. 16.

SCHÖRGENDORFER 558

Celeia

275. Rim, neck, handle, inv.no.: 10404.
s: Stranetora Mislojiceva; red painted inscription on the neck: OL(iva)/ALB(a)/EXD(ulcis)/EXC(ellens); reddish yellow (7.5 YR 7/6), H=29cm, D=14.8cm, s.h.=4.6x2.5cm.
fig. 10.

Poetovio

276. Rim, neck, handles, inv.no.: -,
s: Spodnja Hajdina, st. parc. 95. ko; red painted inscription on the neck: OL(iva)/AL(ba)/EXDUL(cis).
cf. Mikl 1960, 156. T. II. no. 12.

277. Body fr. inv.no.: 10951.
s: Zgornji Breg, parc. st. 293.□ 38-39; reddish yellow (5 YR 7/6), H=17cm, d=25.5cm, s.h.=5.4x2.4cm.
Pl. 12.

278. Rim, neck, handle fr. inv.no.: 1471.
pink (7.5 YR 7/4), H=27.5cm, D=14.3cm, s.h.=5.2x2.4cm.
Pl. 12.

Salla

279. Rim, neck fr. inv.no.: 83.5.30.
cf. Bezeczky 1984, 159. Pl. 6. no. 30.
fig. 27.

280. Rim, neck fr. inv.no.: 83.5.81.
s: Quadrate T - C/II. layer 4-5, area "t"; reddish yellow (7.5 YR 7/6), H=9.3cm, D=14cm.
fig. 27.

281. Neck, body fr. inv.no.: 83.5.82.
s: Quadrate S - A/V at a depth of 250-300cm, from the NW - SE trench, layer 4-5, area "s"; reddish yellow (7.5 YR 8/6), H=7.3cm, d=9.9cm.
fig. 27.

282. Body and handle fr. inv.no.: -,
s: E part of the quadrate T - S/III at a depth of 160-165cm, from brick and sand layer; pink (7.5 YR 7/4), H=6cm.
fig. 27.

283. Handle fr. inv.no.: 83.5.83.
s: From the surface layer, between two buildings, layers 3-4, area "t", pink (7.5 YR 7/4), H=21cm, s.h.=5.3x2.8cm.
fig. 27.

284. Handle fr. inv.no.: 83.5.84.
s: Quadrate S - E/I, from a pit, layers 5-6, area "s"; yellow (10 YR 8/6), H=8cm, s.h.=5.3x2.4cm.
fig. 27.

285. Neck, body fr. inv.no.: -,
s: Quadrate F - D/XV. from the grey charcoal with clay layer at a depth of 40-70cm, under the brick line, layer 4, area "f"; black painted inscription: OLIVA/.../(EX)CEL(lens);
fig. 27.

Savaria

286. Rim, neck, body, base fr. inv.no.: 54.321.26.
reddish yellow (7.5 YR 7/6), H=43.6cm, D=14.2cm, th.r.=1.9cm.
fig. 26.

287. Neck fr., body, base, inv.no.: 54.321.27; 54.321.28; 54.187.102.
s: "Uj street, Spöttel house" today Petőfy street, red painted inscription on the neck: .../EX.(ellens) /..XXII; reddish yellow (7.5 YR 7/6), H=61cm, d=29cm;
cf. Lipp 1875, 87, fig. 36.
fig. 26.

288. Base fr. inv.no.: 65.49.82.
s: Köztársaság square, foundation no. 1 pit no. 3; reddish yellow (5 YR 7/6), H=22.3cm, d=8.6cm.
fig. 26.

289. Rim, neck, handle fr. inv.no.: -,
s: Kossuth street, N from the building at a depth of 50-60cm, red painted inscription on the neck: OL(iva)/NIG(ra)/EXD(ulcis)/EXC(ellens); pale yellow (2.5 YR 7/6), H=17cm, D=11.5cm, th.r.=1.5cm.
fig. 27.

290. Neck, handle fr. inv.no.: 61.3.864.
s: J.P.I.R., NE corner of the building at a depth of 200-240cm; reddish yellow (7.5 YR 7/6), H=12.3cm.
fig. 26.

Carnuntum

291. Rim, neck, handle fr.
cf. Grünewald 1983, T. 44. no. 3.

292. Rim, neck.
cf. Grünewald 1983, T. 44. no. 4.

293. Rim, neck fr.
cf. Grünewald 1983, T. 44. no. 5.

294. Rim, neck fr.
cf. Grünewald 1983, T. 44. no. 6.

Portorecanatian amphora

Emona

295. Rim, neck, handle, inv.no.: -,
s: Rjaia Plast, 92-95, G 77, Ivan Cankar GL 297; reddish yellow (5 YR 7/6), D=14.1cm, th.r.=2.4cm.

296. Neck, body fr. handles, inv.no.: -,
s: Jakopičev vrt. 1964. IP 15/IV-V; reddish yellow (5 YR 6/6).

Poetovio

297. Rim, neck, body fr., handles, inv.no.: 3678.
reddish yellow (5 YR 7/8), H=31cm, D=13.4cm, th.r.=1.7cm, s.h.=3.9x3.6cm.
cf. Mikl Curk 1976, 47, 97.
Pl. 12.

Salla

298. Rim, inv.no.: 83.5.18.
cf. Bezeczky 1984, 158, Pl. 5. no. 18.
fig. 28.

299. Rim, inv.no.: 83.5.19.
cf. Bezeczky 1984, 158, Pl. 5. no. 19.
fig. 28.

300. Rim, inv.no.: 83.5.20.
cf. Bezeczky 1984, 158, Pl. 5. no. 20.
fig. 28.

301. Rim, inv.no.: 83.5.21.
cf. Bezeczky 1984, 159, Pl. 5. no. 21.
fig. 28.

302. Rim, neck, handle, inv.no.: 83.5.22.
cf. Bezeczky 1984, 159, Pl. 5. no. 22.
fig. 28.

303. Rim, inv.no.: 83.4.23.
cf. Bezeczky 1984, 159, Pl. 5. no. 23.
fig. 28.

304. Rim, inv.no.: 83.5.24.
cf. Bezeczky 1984, 159, Pl. 5. no. 24.
fig. 28.

305. Rim, inv.no.: 83.5.25.
cf. Bezeczky 1984, 159, Pl. 5. no. 25.
fig. 28.

306. Rim, neck fr. inv.no.: 83.5.26.
cf. Bezeczky 1984, 159, Pl. 6. no. 26.
fig. 28.

307. Rim, inv.no.: 83.5.27.
cf. Bezeczky 1984, 159, Pl. 6. no. 27.
fig. 28.

308. Rim, inv.no.: 83.5.28.
cf. Bezeczky 1984, 159, Pl. 6. no. 28.
fig. 28.

309. Rim, inv.no.: 83.5.130.
s: from the channel, above the yellow earth at a depth of 150-230cm, layer 4/A-B, area "b"; pink (7.5 YR 7/4), H=7.6cm, D=14.4cm.
fig. 28.

310. Base, inv.no.: 83.5.46.
cf. Bezeczky 1984, 161, Pl. 9. no. 46.
fig. 28.

Savaria

311. Neck, handle fr. inv.no.: 54.91.16; 54.91.26.
s: J.P.I.R.; reddish yellow (5 YR 6/6), H=19.2cm, d=9.3cm, s.h.=3.5x3.4cm.
fig. 28.

NORTH AFRICAN AMPHORA

Poetovio

312. Complete (restored), inv.no.: -,
red (2.5 YR 5/6), H=124cm, D=18.2cm, s.h.=2.4x4.8cm, d=49cm.
Pl. 12.

SPATHEION AMPHORA

Poetovio

313. Complete, inv.no.: 158.
H=94cm, D=11.8cm, d=15.7cm, s.h.=3.2x1.9cm.
Pl. 12.

BÓNIS XXXI/5 amphora

Emona

314. Complete, inv.no.: 6397.
cf. Bónis 1942, XXXI. T. no. 5.

315. Complete (restored), inv.no.: 6764.
cf. Bónis 1942, XXXIII. T. no. 7.

Poetovio

316. Rim, neck, body fr., handles, inv.no.: 3530.
light red (10 R 6/8).

317. Complete, inv.no.: -,
s: Gosilski Dom.

Salla

318. Rim, neck, base, handles, inv.no.: 83.5.32.
cf. Bezeczky 1984, 159. Pl. 7. no. 32.
fig. 30.

319. Rim, inv.no.: 83.5.138.
s: quadrate T - D/II. from the trench S of the wall at a depth of 200-220cm; very pale brown (10 YR 8/4), H=4.8cm, D=11.2cm.
fig. 30.

Savaria

320. Complete, inv.no.: -,
red (2.5 YR 5/8), H=39cm, D=7.7cm, d=17.2cm, s.h.=1.8cm.
fig. 30.

321. Base, inv.no.: 54.91.43.
s: J.P.I.R.; reddish yellow (7.5 YR 7/6), H=7.3cm, d=2.1cm.
fig. 30.

Camulodunum 189 amphora

Carnuntum

322. Rim, neck, body fr.
cf. Grünewald 1983, T. 44. no. 1.

Aegean amphora (Ostia VI, Niederbieber 77, Kapitän II, Zeest 79)

Scarbantia

323. Handle fr. inv.no.: Sc. 7.
s: Town Hall street, Cinema; red (2.5 YR 5/8), H=15.2cm, s.h.=5.6x3.2cm.
fig. 31.

Brukner 167/117 amphora base

Poetovio

324. Base, inv.no.: 3691.
light red (10 R 6/6), H=24cm, d=10.2cm.

Savaria

325. Base, inv.no.: 64.15.2515.
s: Alkotmány street, quadrate nos. 107-108 at a depth of 210-270cm, from brown debris layer; pink (5 YR 7/4), H=22.6cm.
fig. 21.

MISCELLANEOUS LARGE AMPHORAE

Emona

326. Body fr., base, inv.no.: -,
s: Jakopičev vrt. at a depth of 100cm; H=14cm.

Scarbantia

327. Body fr. base, inv.no.: 79.2.97; 79.2.180.
s: Arpád street, from area of IKVA department store, stray find beside the kiln; light red (10 R 6/8), H=22cm.
fig. 35.

MAGDALENSBERG 1958/fig.25 no. 10

Emona

328. Rim, inv.no.: F 1693.

Salla

329. Rim, inv.no.: 83.5.128.
s: Quadrat G/XVI, from the great channel at a depth of 120-130cm, from the burnt grey layer, from the NE-SW channel, layers 3-4/A, area "d"; reddish yellow (5 Y 6/6), H=5cm, D=16cm.
fig. 29.

MISCELLANEOUS AMPHORAE

Poetovio

330. Body fr., base, inv.no.: 2193.
H=34.5cm, d=27.5cm.
cf. Mikl Curk 1976, 34, 87.
Pl. 12.

331. Body fr., base, inv.no.: 3542.
s: Zgornja Hajdina, Maruškova Njiva; H=46cm, d=26.3cm.
cf. Mikl Curk 1976, 45, 96.
Pl. 12.

332. Body fr., base, inv.no.: 10950.
s: Zgornji Breg, Parc. st. 293. 41; H=36cm, d=5.6cm.

Salla

333. Rim, inv.no.: 83.5.140.
s: Room no. I beside the N wall, above the floor at a depth of 150cm; light red (2.5 YR 6/6), H=4cm, D=11.2cm.
fig. 29.

334. Rim, inv.no.: 83.5.137.
s: stray find; reddish yellow (5 YR 6/6), H=4.8cm, D=12cm.
fig. 29.

335. Rim, inv.no.: 83.5.139.
s: stray find; pink (7.5 YR 8/4), H=4.4cm, D=14.4cm.
fig. 29.

336. Rim, inv.no.: 83.5.144.
s: N part of F/I. from brown earth at a depth of 160cm; pink (7/5 YR 7/4), H=5.2cm, D=11.6cm.
fig. 29.

337. Neck fr. inv.no.: 83.5.33.
cf. Bezeczky 1984, 160, Pl. 7. no. 33.
fig. 31.

338. Neck fr. inv.no.: 83.5.34.
cf. Bezeczky 1984, 160. Pl. 7. no. 34.
fig. 31.

339. Neck fr. inv.no.: 83.5.35.
cf. Bezeczky 1984, 160. Pl. 7. no. 35.
fig. 31.

340. Neck fr. inv.no.: 83.5.36.
cf. Bezeczky 1984, 160. Pl. 7. no. 36.
fig. 31.

341. Handle fr. inv.no.: 83.5.126.
s: Quadrate C/VII at a depth of 185cm, layers 5-6, area "c"; light red (2/5 YR 6/6), H=11cm, s.h.=3.1cm.
fig. 32.

342. Handle fr. inv.no.: 83.5.127.
s: From the S part of the quadrate A/VI, from the gravel layer at a depth of 100-140cm, layer 4/C - 5, area "c"; reddish yellow (5 YR 6/6), H=10.1cm, s.h.=3.2cm.
fig. 32.

343. Handle fr. inv.no.: 83.5.121.
s: Quadrate K 3, from the wall demolition layer at a depth of 50-150cm, layer 4/D - 5, area "j"; light brown (5 YR 6/4), H=10.7cm, s.h.=3.2x2.7cm.
fig. 32.

344. Handle fr. inv.no.: 83.5.143.
s: S corner of room no. I, 0-80cm below the brick line layer 4/D, area "c"; pink (7.5 YR 7/4), H=6.4cm, s.h.=2.4x3.2cm.
fig. 32.

345. Handle fr. inv.no.: 83.5.134.
s: Quadrate K 2, from clay and sand layer at a depth of 140-160cm, layer 4/D - 5, area "j"; pink (5 YR 7/4), H=10.2cm, s.h.=4.1x2.5cm.
fig. 32.

346. Handle fr. inv.no.: 83.5.65.
cf. Bezeczky 1984, 162. Pl. 11. no. 65.
fig. 32.

347. Handle fr. inv.no.: 83.5.66.
cf. Bezeczky 1984, 162. Pl. 11. no. 66.
fig. 32.

348. Handle fr. inv.no.: 83.5.67.
cf. Bezeczky 1984, 162. Pl. 12. no. 67.
fig. 32.

349. Handle fr. inv.no.: 83.5.120.
s: Post Office Garden, below the yard and from the brick level, layer 1/B - 3, area "d"; reddish yellow (5 YR 7/8), H=15.3cm, s.h.=3.6x3.3cm.
fig. 33.

350. Handle fr. inv.no.: 83.5.123.
s: Quadrate E/III, from the yellow layer at a depth of 80-140cm, layer 4, area "h"; reddish yellow (5 YR 6/6), H=11.5cm, s.h.=3.4x3.2cm.
fig. 33.

351. Handle fr. inv.no.: 83.5.122.
s: Quadrate - A/III, under the wall, stray find, layers 4-5, area "b"; reddish yellow (5 YR 6/6), H=14cm, s.h.=3.2cm.
fig. 33.

352. Handle fr. inv.no.: 83.5.125.
s: From the NW part of quadrate E/II, alongside the kiln at a depth of 210cm, layers 5-6, area "n"; light red (2.5 YR 6/6), H=12cm, s.h.=3.1cm.
fig. 33.

353. Handle fr. inv.no.: 83.5.124.
s: NW part of room I, from the wall at a depth of 0-30cm, layers 3-4/B, area "d"; red (2.5 YR 5/6), H=9.3cm, s.h.=3.7x3.4cm.
fig. 33.

354. Handle fr. inv.no.: 83.5.68.
cf. Bezeczky 1984, 163. Pl. 12. no. 68.
fig. 33.

355. Handle fr. inv.no.: 83.5.64.
cf. Bezeczky 1984, 162. Pl. 11. no. 64.
fig. 33.

356. Handle fr. inv.no.: 83.5.69.
cf. Bezeczky 1984, 163. Pl. 12. no. 69.
fig. 33.

357. Handle fr. inv.no.: 83.5.70.
cf. Bezeczky 1984, 163. Pl. 12. no. 70.
fig. 32.

358. Neck, body fr. inv.no.: 83.5.154.
s: NE corner of room no. II, 60cm under the brick line, layer 4/B, area "c"; light red (2.5 YR 6/6), H=9.5cm.
fig. 31.

359. Neck, body fr. inv.no.: 83.5.142.
s: Room no. I, 90cm under the brick line, layer 4/D, area "c"; light grey (5 YR 7/2), H=12cm.
fig. 22.

360. Handle fr. inv.no.: 83.5.85.
s: Quadrate S - B/II at a depth of 180-190cm; pink (5 YR 8/4), H=15cm, s.h.=4.9x2.8cm.

361. Handle fr. inv.no.: 83.5.145.
s: Room no. I, 0-50cm under the brick line, layer 4/B, area "c"; red (2.5 YR 5/8), H=5.3cm.
fig. 31.

362. Base, inv.no.: 83.5.39.
cf. Bezeczky 1984, 160. Pl. 8. no. 39.
fig. 35.

363. Base, inv.no.: 83.5.40.
cf. Bezeczky 1984, 160. Pl. 8. no. 40.
fig. 35.

364. Base, inv.no.: 83.5.42.
cf. Bezeczky 1984, 160. Pl. 8. no. 42.
fig. 35.

365. Base, inv.no.: 83.5.41.
cf. Bezeczky 1984, 160. Pl. 8. no. 41.
fig. 35.

366. Base, inv.no.: 83.5.43.
cf. Bezeczky 1984, 160. Pl. 8. no. 43.
fig. 35.

367. Base, inv.no.: 83.5.131.
s: Quadrate S - C/I, from NW-SE wall at a depth of 90-170cm, layer 1 - 4, area "s"; reddish yellow (7.5 YR 7/6), H=6.6cm, d=3cm.
fig. 35.

368. Rim, neck, body fr. inv.no.: 83.5.31.
cf. Bezeczky 1984, 159. Pl. 7. no. 31.
fig. 30.

369. Base, inv.no.: 83.5.38.
cf. Bezeczky 1984, 160. Pl. 8. no. 38.
fig. 35.

Savaria

370. Rim, inv.no.: 54.411.10.
s: J.P.I.R., at a depth of 400cm, SE part of the apse; pink (7.5 YR 7/4), H=5.9cm, D=12cm, th.r.=1.3cm.
fig. 18.

371. Rim, inv.no.: 69.1.149.
s: Sanctuary of Isis, quadrate no. 1624 at a depth of 130-175cm; light red (2.5 YR 6/8), H=7.4cm, D=13.6cm, th.r.=2.2cm.
fig. 18.

372. Rim, inv.no.: 66.3.113.
s: R.C., trench no. 10-11, on the E, outside the wall at a depth of 190-300cm; reddish yellow (5 YR 6/8), H=5.1cm, D=13.5cm, th.r.=2.1cm.
fig. 18.

373. Rim, inv.no.: 69.1.4296.
s: Sanctuary of Isis, quadrate no. 1972 at a depth of 130-175cm; light red (2/5 YR 6/8), H=8cm, D=14cm, th.r.=1.8cm.
fig. 18.

374. Rim, neck, handle fr., inv.no.: 75.1.817.
s: J.P.I.R.; quadrates nos. 2/C-B; reddish yellow (5 YR 6/6), H=26.5cm, D=6.6cm, s.h.=4x1.5cm.
fig. 29.

375. Body, handle fr., inv.no.: 74.4.1589.
s: Szt. Márton street, N part of trench no. 3 at a depth of 200-210cm, brown charcoal layer; reddish yellow (5 YR 7/6), H=24.5cm.
fig. 30.

376. Rim, handle fr. inv.no.: 54.91.4.
s: J.P.I.R.; red (2/5 YR 5/8), H=10.7cm, D=11cm, th.r.=1.4cm, s.h.=3cm.
fig. 30.

377. Neck, handle fr. inv.no.: 64.15.3599.
s: Alkotmány street, NE part of part quadrate no. 146-166 at a depth of 290-330cm; yellowish red (5 YR 5/6), H=9.8cm.
fig. 31.

378. Neck fr. inv.no.: 64.15.2086.
s: Alkotmány street, S part of the quadrate no. 103 at a depth of 90-220cm, from brick mortar filling; reddish yellow (5 YR 6/8), H=11.5cm.
fig. 31.

379. Handle fr. inv.no.: 54.91.25.
s: J.P.I.R.; light red (2.5 YR 6/8), H=15cm, s.h.=3.2x3.5cm.
fig. 32.

380. Handle fr. inv.no.: 66.3.1637.
s: R.C., trench no. 16 at a depth of 160-260cm; light red (2.5 YR 6/8), H=13.2cm, s.h.=4cm.
fig. 32.

381. Handle fr. inv.no.: 61.3.865.
s: J.P.I.R.; NE corner of the building at a depth of 200-240cm; reddish yellow (7.5 YR 7/6), H=22cm, s.h.=3.3x3.6cm.
fig. 33.

382. Handle fr. inv.no.: 58.1.334.
s: Thököly street, quadrate no. 103 at a depth of 190cm, from debris; reddish yellow (5 YR 6/6), H=8cm, s.h.=4.2x3.3cm.
fig. 33.

383. Handle fr. inv.no.: 54.91.21.
s: J.P.I.R.; reddish yellow (7.5 YR 6/6), H=16.8cm, s.h.=3.4x3.2xcm.
fig. 33.

384. Handle fr. inv.no.: 57.1.983.
s: Thököly street, outside quadrate no. 65, from the gravel with lime layer; light red (2.5 YR 6/8), H=16.8cm, s.h.=3.4x3.3cm.
fig. 33.

385. Handle fr. inv.no.: 59.1.699.
s: Thököly street, E part of quadrate no. 403 at a depth of 110-160cm, between two gravel layers; light red (2.5 YR 6/8), H=10.4cm, s.h.=3.7x3.6cm.
fig. 33.

386. Handle fr. inv.no.: 64.15.390.
s: Alkotmány street, quadrate no. 65, outside the N wall at a depth of 180-220cm; reddish yellow (5 YR 7/6), H=10cm, s.h.=3.5x2.9cm.
fig. 33.

387. Base, inv.no.: 77.2.2036.
s: Town Hall, SW part of quadrate no. 4, layer no. 132 at a depth of 250-270cm, from red burnt layer intermixed with charcoal; red (2.5 YR 5/8), H=15.5cm, d=2.8cm.
fig. 34.

388. Base, inv.no.: 54.186.5.
s: From the site of the Palace Hotel; red (2.5 YR 5/8), H=12.6cm, d=2.7cm.
fig. 34.

389. Base, inv.no.: 64.15.2221.
s: Alkotmány street, quadrate no. 88 at a depth of 280-305cm, from black clay soil; reddish yellow (5 YR 6/6), H=10.5cm, d=2.6cm.
fig. 34.

390. Base, inv.no.: 80.1.1966.
s: Sörház street, quadrate no. 83 at a depth of 240-290cm, from disturbed debris layer; white (10 YR 8/2), H=9.3cm, d=4cm.
fig. 34.

391. Base, inv.no.: 64.14.270.
s: Sanctuary of Isis, quadrate no. 57, outside the concrete foundation; red (2.5 YR 5/6), H=10.6cm, d=2.7cm.
fig. 34.

392. Base, inv.no.: 66.3.1671.
s: R.C., trench no. 3 at a depth of 130-180cm, from N of the wall, from burnt clay; reddish yellow (7.5 YR 5/6), H=10.5cm, d=4.3cm.
fig. 34.

393. Base, inv.no.: 80.1.3798.
s: Sörház street, quadrate no. 72 at a depth of 260-280cm, from burnt charcoal layer; reddish yellow (7.5 YR 6/6), H=10.9cm, d=4.9cm.
fig. 34.

394. Base inv.no.: 54.91.38.
s: J.P.I.R.; red (2.5 YR 4/6), H=11.8cm, d=5.3cm.
fig. 34.

395. Base, inv.no.: 54.91.42.
s: J.P.I.R.; very pale brown (10 YR 7/4), H=9.5cm, d=3.8cm.
fig. 35.

396. Base, inv.no.: 70.1.924.
s: Sanctuary of Isis, quadrate no. 2670-2690. debris layer; very pale brown (10 YR 7/4), H=9.3cm, d=3cm.
fig. 35.

397. Base, inv.no.: 68.27.14.
s: Thököly street, ÉDÁSZ building at a depth of 400cm, from the flooring slabs; light red (2/5 YR 6/8), H=10.3cm, d=3.8cm.
fig. 35.

398. Base, inv.no.: 65.50.335.
s: Ady square, Coach station, from a pit; very pale brown (10 YR 8/4), H=14.5cm, d=4cm.
fig. 35.

399. Base, inv.no.: 54.91.44.
s: J.P.I.R.; light red (2.5 YR 6/8), H=4.5cm, d=2.9cm.
fig. 35.

400. Base, inv.no.: 62.9.300.
s: Thököly street, quadrate no. 179-213 at a depth of 220-260cm, from yellowish clay layer; red (2.5 YR 5/6), H=14.2cm, d=3.3cm.
fig. 35.

401. Base, inv.no.: 54.91.37.
s: J.P.I.R.; reddish brown (2.5 YR 4/4), H=11.8cm, d=3.4cm.
fig. 35.

402. Base, inv.no.: 67.1.67.
s: In the courtyard of the Bishop's palace, quadrate no. 3 at a depth of 400cm, E side of the NS wall; red (10 YR 5/6), H=9.7cm, d=2.8cm.
fig. 35.

403. Base, inv.no.: 66.3.264.
s: R.C., E part of trench no. 13; light reddish brown (5 YR 6/4), H=8.3cm, d=2.8cm.
fig. 35.

404. Base, inv.no.: 54.91.36.
s: J.P.I.R.; light brown (7.5 YR 6/4), H=12.7cm, d=2.8cm.
fig. 35.

405. Base, inv.no.: 77.2.5538.
s: Town Hall, quadrate no. 13 at a depth of 200-230cm, above the burnt soil layer; yellowish red (5 YR 5/8), H=9.4cm, d=4.5cm.
fig. 35.

406. Base, inv.no.: 54.431.4.
s: J.P.I.R.; reddish yellow (7.5 YR 6/6), H=11.4cm, d=5.8cm.
fig. 35.

407. Base, inv.no.: 65.49.198; 65.49.188.
s: Köztársaság square, foundation trench no. 1, pit no. 2 at a depth of 350cm, from the fill; red (2.5 YR 7/4), H=10.1cm, d=5cm.
fig. 35.

Scarbantia

408. Rim, neck, handle fr. inv.no.: 55.111.27.
s: Town Hall; light red (2.5 YR 6/6), H=12.8cm, D=11.8cm, th.r.=1.8cm.
fig. 29.

409. Rim, neck, body fr. handles, inv.no.: 55.111.21.
s: Town Hall; reddish yellow (7.5 YR 6/6), H=21cm, D=13.5cm, th.r.=2.4cm.
fig. 30.

410. Neck, body fr. inv.no.: Sc. 9.
s: Új and Szt. György streets, from the NE corner of the cellar at a depth of 382cm; red (10 R 5/8), H=16cm.
fig. 31.

411. Neck fr. inv.no.: Sc. 5.
s: Town Hall, from filling of the pit; light red (10 R 6/8), H=8.9cm.
fig. 31.

412. Base, inv.no.: 64.24.1.
s: Fabricius house, at the city wall from a depth of 880-900cm, above the green clay; reddish yellow (7.5 YR 7/6), H=13cm.
cf. Holl-Nováki-Póczy 1962, 51.
fig. 34.

413. Base, inv.no.: Sc. 3.b.
s: Town Hall, from the green layer under the stone dust level; very pale brown (10 YR 8/4), H=7.6cm.
fig. 34.

414. Base, inv.no.: Sc. 2.
s: Városház street, below the stone dust layer; very pale brown (10 YR 7/4), H=14.5cm.
fig. 35.

415. Base, inv.no.: 55.111.29.
s: Town Hall; pink (7.5 YR 7/4), H=10.8cm.
fig. 35.

416. Base, inv.no.: Sc. 1.
s: Town Hall street, below the stone dust layer; reddish yellow (7.5 YR 8/6), H=12.4cm.
fig. 35.

Stoppers

Salla

417. inv.no.: 83.5.71.
cf. Bezeczky 1984, 163. Pl. 12. no. 71.
fig. 36.

418. inv.no.: 83.5.72.
cf. Bezeczky 1984, 163. Pl. 12. no. 72.
fig. 36.

419. inv.no.: 83.5.73.
cf. Bezeczky 1984, 163. Pl. 12. no. 73.
fig. 36.

420. inv.no.: 83.5.76.
cf. Bezeczky 1984, 163. Pl. 12. no. 76.
fig. 36.

421. inv.no.: 83.5.74.
cf. Bezeczky 1984, 163. Pl. 12. no. 74.
fig. 36.

422. inv.no.: 83.5.75.
cf. Bezeczky 1984, 163. Pl. 12. no. 75.
fig. 36.

423. inv.no.: 83.5.99.
s: Quadrate - A/IV, layers 5-6, from the grey gravel layer, area "b"; reddish yellow (5 YR 6/6), H=7.7cm, d=5.9cm.
fig. 36.

424. inv.no.: 83.5.98.
s: Quadrate T - B/IV, N from the NE-SW wall in the layer of charcoal and clay at a depth of 200-230cm, layer 5, area "t"; light red (10 R 6/8), H=9.4cm, d=6.1cm.
fig. 36.

Savaria

425. inv.no.: 54.321.31.
red (2.5 YR 5/8), H=12.7cm, d=7.7cm.
fig. 36.

426. inv.no.: 54.222.2.
brown (7.5 YR 5/4), H=7.3cm, d=4.9cm.
fig. 36.

427. inv.no.: 66.3.3147.
reddish yellow (5 YR 7/6), H=8.3cm, d=5.2cm.
fig. 36.

428. inv.no.: 54.91.57.
light red (2.5 YR 6/8), H=5.1cm.
fig. 36.

429. inv.no.: 54.431.5.
very pale brown (10 YR 7/3), H=8.9cm, d=4.9cm.
fig. 36.

430. inv.no.: 54.91.53.
reddish yellow (5 YR 7/8), H=6.7cm, d=4.9cm.
fig. 36.

431. inv.no.: 65.49.61.
light red (2.5 YR 6/8), H=9.1cm, d=5cm.
fig. 36.

432. inv.no.: 76.1.53.
light red (10 YR 6/8), H=6cm, d=5.5cm.
fig. 36.

433. inv.no.: 54.91.51.
pale yellow (2.5 Y 8/4), H=6.3cm, d=5.4cm.
fig. 36.

434. inv.no.: 54.91.49.
light red (2.5 YR 6/6), H=6.3cm, d=6.2cm.
fig. 36.

435. inv.no.: 54.91.54.
light red (2.5 YR 6/8), H=4.8cm, d=5cm.
fig. 36.

436. inv.no.: 54.91.48.
light red (2.5 YR 6/8), H=6.8cm, d=4.9cm.
fig. 36.

437. inv.no.: 54.187.55.
light red (2.5 YR 6/6), H=9.3cm, d=6.5cm.
fig. 36.

438. inv.no.: 66.3.610.
pink (5 YR 8/4), H=11cm, d=7.6cm.
fig. 36.

439. inv.no.: 54.91.55.
light red (2.5 YR 6/8), H=4.4cm.
fig. 36.

440. inv.no.: 54.91.50.
pink (7/5 YR 8/4), H=4.8cm.
fig. 36.

441. inv.no.: 54.187.142.
reddish yellow (7/5 YR 7/6), H=6.8cm, d=5.4cm.
fig. 36.

442. inv.no.: 54.187.56.
light red (2/5 YR 6/8), H=6.3cm, d=5.6cm.
fig. 36.

Scarbantia

443. inv.no.: 52.234.1.
reddish yellow (5 YR 7/8), H=7.7cm, d=5.1cm.
fig. 36.

BIBLIOGRAPHY

Abramič, M. 1925. Poetovio, Führer durch die Dankmäler der Römischen Stadt Wien.

Alzinger, W. 1955. Kleinfunde von Aguntum aus den Jahren 1950 bis 1952 Wien.

Baldacci, P. 1967. 'Alcuni aspetti dei commerci nei territori Cisalpini' in Centro Studi e Documentazione sull'Italia Romana, Atti 1. Milans - Varese 1967-68, 7-50.

Baldacci, P. 1969. 'Le principali correnti del commerci di anfore romane nella Cisalpina dal III. sec. a.C. al II. d.C.' In Problemi della Ceramica Romana di Ravenna, dalla Valle Padana e dell'Alto Adriatico (Atti del Convegno Internazionale di Ravenna), 103-31. Bologna.

Baldacci, P. 1972. 'Importazioni Cisalpine e produzione Apula'. In Recherches sur les Amphores Romaines (Collection de l'Ecole Française de Rome, No. 10) 7-28 Rome.

Barkóczi, L. & Bónis, É. 1954. Das frührömische Lager und die Wohnsiedlung von Adony (Vetus Salina), Acta Arch. Hung., 4, 129-

Beltrán Lloris, M. 1970. Las Anforas Romanas en Espana (Monografias Arquelógicas, Anejos de Caesaraugusta, No. 8) Zaragoza.

Bezeczky, T. 1984. 'Roman amphorae from Zalalövő' Mitteilungen des Archäeologischen Instituts der Ungarischen Akademie der Wissenschaften 12/13, 1982/83, 153-166. Pls. 1-12.

Bezeczky, T. 1985. 'P. Iturius S. amphorája Vetus Salinában' Folia Archaeologica, 36, 69-74.

Bónis, É. 1942. Die kaiserzeitliche Keramik von Pannonien I. (Dissertationes Pannonicae, II. 20) Budapest.

Bónis, É. 1969. Die spatkeltische Siedlung Gellérthegy-Tabán in Budapest Budapest.

Bonsor, G.E. 1931. The Archaeological Expedition along the Guadalquivir, 1889-1901 New York.

Bohn, O. 1926. 'Pilsenschriften auf Amphoren aus Augst und Windisch' Anzeiger für Schweizerische Altertumskunde 28.4, 197-212.

Bojovič, D. 1977. Rimska keramika Singidunuma Beograd.

Bruckner, A. 1966. 'Ziegel und Amphorenstempel aus Chur'. in Neue Römische Inschriften und Kleinfunde aus dem Engadin und aus Chur (Schriftenreihe des Ratischen Museums Chur. Heft 2).

Brukner, O. 1981. Rimska keramika u Jugoslovenskom delu provincije donje Panonije (Dissertationes et Monographie No. 24) Beograd.

Buchi, E. 1973. 'Banchi di anfore romaine a Verona. Note sui commerci Cisalpini, "Il territotio Veronese in età Romana"'. In Atti del Covegno tenuto a Verona 1971, 531-637. Verona.

Buchi, E. 1974. 'Commerci delle anfore "istriane"' Aquileia Nostra 45/46 for 1974-75, 431-444.

Buocz, T. 1961. 'Korarómai sirok a szombathelyi Rumi úton' Archaeologiai Értesitő, 81, 219-239.

Boltin-Tome, E. 1974. 'Žigi na Rimskih opekah iz depoja Pomorskega Muzeja "Sergej Mašera" v Piranu' Arheološki Vestnik, 25, 225-232.

Callender, M.H. 1965. Roman Amphorae with Index of Stamps London.

Carre, M.B. 1985. 'Les amphores de la Cisalpine et de l'Adriatique au début de l'Empire' MEFRA, 97, 207-45.

Csermémyi, V. & Tóth, E. 1979. 'Eine römische Strassenstation und die Strassenstrecke zwischen Salla und Arrabona, 'SAVARIA' A Vas Megyei Múzeumok Értesitője, 13/14 for 1979-1980, 171-203.

Degrassi, A. 1942. 'Un nuovo frammento dei Fasti dei Sodales Augustales Claudiales' Epigraphica, 4, 17-22.

Degrassi, A. 1953. 'Aquileia e l'Istria in eta Romana'. In Studi Aquileiesi offerti a G. Brusin, 51-65 Aquileia.

Degrassi, A. 1956. 'L'Esportazione di olio e olive istriane nell'eta Romana'. In Atti e Memorie della Societa Istriana di archaeologia e storia patria, 104-112.

Degrassi, N. 1951. 'Milano, Via Rugabella' Notizie degli Scavi di Antichita, 76. 49-50.

Dressel, H. 1879. 'Di un grande deposito di anfore rinvenuto nel nuovo quartiere del Castro Pretorio' Bulletino della Commissione Archaeologica Comunale di Roma, 7, 36-112; 143-196.

Ettlinger, E. 1977. 'Aspects of amphora typology seen from the North'. In Méthodes Classiques et Methodes Formelles dans l'Etude des Amphores (Collection de l'Ecole Francaise de Rome, No. 32), 9-16. Roma.

Frova, A. 1952. 'Marche di anfore e altri bolli romani del Milanese' *Epigraphica*, 14, 49-93.

Fuchs, M. 1978. *Die römischen Amphoren von Magdalensberg, Band I-III* (Dissertation, Universitat Innsbruck).

Gábler, D. & Kelemen, M. 1984. 'Olio betico in Pannonia. Anfore ispaniche nella Valle Danubiana' *Archivo Espanol de Arqueologia*, 57, 121-42.

Gnirs, A. 1904. 'Antike Funde aus Pola und Umgebung' *JÖAI*, 7. beibl. 131-46.

Gnirs, A. 1910A. 'Eine antik-römische Tonwarenfabrik und ihr Warendepot bei Pola' *JÖAI*, 13, 95-103.

Gnirs, A. 1910B. 'Eine römische Tonwarenfabrik in Fasana bei Pola' *Jahrbuch für Altertumskunde*, 4, 79-88.

Gnirs, A. 1911. 'Fasana: Neue Funde aus der Figlina des C. Laecanius Bassus' *JÖAI*, 14, beibl. 35-39.

Gnirs, A. 1914. 'Forschungen in Pola und in der Polesana' *JÖAI*, 17, beibl. 161-84.

Grace, V.R. 1961. *Amphoras and the ancient wine trade*, (American School of Classical Studies at Athens, Excavation of the Athenian Agora, No. 6) Princeton.

Gregorutti, C. 1986. 'La figulina imperiale Pansiana di Aquileia ed i prodotti fittili dell'Istria'. In *Atti e Memorie della Societa Istriana di Archeologia e Storia Patria*, 2, fasc. I-II, 1-37.

Grünewald, M. 1983. Die Funde aus dem Schutthügel des Legionslagers von Carnuntum *Der Römische Limes in Österreich*, 32, 34-35. T. 44-45.

Guenoche, A. & Hesnard, A. 1983. 'Typologie d'amphores romaines par une methode logique de classification' *Computers and the Humanities*, 17, 185-98.

Hamon, E. & Hesnard, A. 1977. 'Problemes de documentation et de description relatifs a un corpus d' amphores romaines'. In *Methodes Classiques et Methodes Formelles dans l'Etude des Amphores* (Collection de L'Ecole Francaise de Rome, No. 32) 17-33. Rome.

Hawkes, C.F.C. & Hull, M.R. 1947. *Camulodunum* (Reports of the Research Committee of Society of Antiquaries of London, No. 14) London.

Hesnard, A. 1980. 'Un dépôt Augustéen d'amphores à la Longarina, Ostie' in F.H. D'Arms & E.C. Kopff (eds.) The Seaborne Commerce of Ancient Rome: Studies in Archaeology and History Memoirs of the American Academy in Rome, 36, 141-56.

Holl, I., Nováki, Gy., & Póczy, K. 1962. 'Városfalmaradványok a soproni Fabricius-ház alatt' Archaeologiai Ertesito 89, 47-67.

Jobst, W. & Piccottini, G. 1986. 'Die Inschriften 1972 bis 1976 und 1978 in Magdalensberg' in Bericht 15, H. Vellers - G. Piccottini: Die Ausgrafungen auf deur Magdalensberg 1975-1979, Klagenfurt, 227-77.

Kapitän, G. 1961. 'Schiffsfrachten antiker Baugesteine und Architekturteile vor den Küsten Ostsiziliens' Klio, 276-318.

Keay, S.J. 1984. Late Roman Amphorae in the Western Mediterranean A Typology and Economic Study: the Catalan Evidence (British Archaeological Reports, International Series, No. 196) Oxford.

Kenner, F. 1911. 'Römische Funde aus Wien 1909-1910' Jahrbuch für Altertumskunde, 5.

Kenner, H. 1947. In C. Praschniker Der Bäderbezirk von Virunum Wien.

Kubitschek, W. & Frankfurter, 1923. S. Führer durch Carnuntum Wien.

Lamboglia, N. 1955. 'Sulla Cronologia delle anfore romane di eta republicana (II-I. secolo a.C)' Rivista Studi Liguri 21, 241-270.

Lipp, V. 1875. 'A római agyagipar a Vasm. Régiségtárban' Vasmegyei Régészeti Egylet Evi Jelentése 62-89.

Medgyes, M. 1977. 'Leletmentések Savaria E-i temetőjében 1970-75' SAVARIA: A Vas Megyei Muzeumok Ertesitője, 11/12 for 1977-78, 177-208.

Mercando, L. 1974. 'La necropoli romana di Portorecanati' Notizie degli Scavi di Antichita, 28, 145-

Mercando, L. 1979. 'Portorecanati (Macerata)' Notizie degli Scavi di Antichita, 33, 180-280.

Mikl, I. 1960. 'Rimske najdbe iz Ptuja po letu 1954' Archeoloski Vestnik for 1960-61, 11-12 153-186.

Mikl-Curk, I. 1976. Poetovio I Katalogi in Monografije No. 13. Ljubljana.

Mócsy, A. 1980. 'Osztrigahéj Zalalövorol' Archaeologiai Ertesitö, 107, 79-80.

Nagy, T. 1948. 'Az albertfalvai római telep' Antuquitas Hungarica, 2, 92-114.

Neumann, A. 1967. Forschungen in Vindobona 1948 bis 1967. I. Teil. Lager und Lagerterritorium (Der Römische Limes in Österreich, No. 23). Wien.

Oelmann, F. 1914. Die Keramik des Kastells Niederbieber Frankfurt a.M.

Pais, H. 1884. Corpus Inscriptionum Latinarum (Supplementa Italica), 156-167.

Panella, C. 1970. 'Ostia, Terme del Nuotatore Ambiente I, Strato V. Le anfore' Studi Miscellanei, 16 (Ostia II), 102-156.

Panella, C. 1974. 'Appunti su un Gruppo di Anfore della Prima, Media e Tarda Eta Imperiale /Secoli I-V. Studi Miscellanei 21 (Ostia III), 460-633.

Peacock, D.P.S. 1977. 'Roman amphorae: typology, fabric and origins'. In Méthodes Classiques et Mèthodes Formelles dans l'Etude des Amphores (Collection de l'Ecole Francaise de Rome, No. 32), 261-78. Rome.

Pélichet, E. 1946. 'A propos des amphores romaines trouvées a Nyon' Zeitschrift für Schweizerische Archäologie und Kunstgeschichte, 8, 189-209.

Plesnicar-Gec, L. 1972. Severno Emonsko Grobisce Katalogi in Monografije No. 8. Ljubljana.

Plesnicar-Gec, L. 1983. Starokrscanski Center v Emoni Katalogi in Monografije No. 21. Ljubljana.

Póczy, K. 1956. 'Római épületek Obudán a Kiscelli u. 10. alatt' Budapest Régiségei, 16, 42-

Riley, J.A. 1982. 'The coarse Pottery from Berenice'. In J. Lloyd Excavations at Sidi Khrebish Benghazi (Berenica) vol. 2 (Suppl. Libya Antiqua), 91-466.

Schörgendorfer, A. 1942. Die Römerzeitliche Keramik der Ostalpenländer Wien - Brünn - München.

Schallmayer, E. 1983. 'Römische Okkupationslinien in Obergermanien und Raetien, Zur chronologischen Typologie der Amphoren'. In Produccion y Commercio del Aceite en la Antigüedad, Segundo Congrese Internacional, 281-336 Madrid.

Schönberger, H. 1978. Kastell Oberstimm (Limesforschungen, No. 18) Berlin.

Szőnyi, E. 1981. 'Kutatások Mursellában' Communicationes Archaeologicae Hungariae, 87-119.

Tassaux, F. 1982. 'Laecanii' Recherches sur une famille senatoriale D'Istrie' MEFRA 94, 227-269.

Tassaux, F. 1983. 'L'Implantation territoriale des grandes familles d'Istrie sous le Haut-Empire Romain' Atti dei civici Musei di Storria ed Arte di Trieste, Quaderno 13.2, for 1983/84, 193-229.

Tchernia, A. 1967. 'Les amphores romaines et l'histoire économique' Journal des Savants, (Oktober/December 1967) 216-34.

Tchernia, A. 1971. 'Les amphores vinaires de Tarraconaise et leur exportation au debut de l'Empire' Archivo Espanol de Arqueologia, 44, 38-85.

Tchernia, A. & Zevi, F. 1972. 'Amphores vinaires de Campanie et de Tarraconaise à Ostie'. In Recherches sur les Amphores Romaines (Collection de l'Ecole Francaise de Rome No. 10.) 35-67. Rome.

Ulbert, G. 1959. Die Römischen Donau-Kastelle Aislingen und Burghöfe (Limesforschungen, No. 1) Berlin.

Vrsalovič, D. 1974. Istraživanja i Zaštita Podmorskih Archeoloskih spomenika u SR Hrvatskoj Zagreb.

Zevi, F. 1966. 'Apunti sulle anfore romane' Archeologia Classica, 18, 208-247.

Zevi, F. 1967. 'Anfore Istriane Ostia' Atti e Memorie della Societa Istriana di Archeologia e Storia Patria, 21-31.

Zeest, I.B. 1960. Keramicheskaja Tara Bospora Moscow.

NOTES

1. We mention only some of the works published: A. Mócsy Pannonia and Upper Moesia, London 1974; idem Die Bevölkerung von Pannonien bis zu den Markomannen-krigen, Budapest, 1959; E. Tóth 'Die probleme der Geschichte Pannoniens' Studia Antiqua, 23, (1976); idem 'Zur Entstehung der gemeinsamen Grenzen zwischen Pannonien und Noricum' Arheološki Vestnik, 31, 1980.80- ; D. Gábler Italische Sigillaten in Nordwest-Pannonien; Wiss Arbeiten aus dem Burgenland 51 Eisenstadt 1973.

2. A. Mócsy outlined some of the problems in connection with the import of foodstuffs: PWRE, 9, Suppl., 681-682. Amphorae published with other finds along the Amber Route: Plesničar-Gec 1972 and 1983; Mikl 1960; Abramič 1925; Mikl-Curk 1976; Csereményi & Tóth 1979; Buócz 1961; Mócsy 1954; Kubitschek & Frankfurter 1923, etc.

3. Dressel 1879 and C.I.L. XV.2.

4. The most significant proposals can be found in the following works: Pélichet 1946; Lamboglia 1955 and Beltrán Lloris 1970. But in most works only one type of amphorae is described in detail: Baldacci 1967, Tchernia 1971, Buchi 1973, Tchernia & Zevi 1972 etc.

5. Baldacci 1967; Baldacci 1969 and Baldacci 1972.

6. Buchi 1973 and 1974.

7. Hamon & Hesnard 1977.

8. Pál Kerékfy and Lajos Rónyai participated in this research.

9. Petrological analyses: Supplement 1.

10. Márta Balla and János Bérczi participated in this research.

11. The Carnuntum finds are known from exhibitions at the Museum Carnuntinum, and from the periodical Carnuntum Jahrbuch.

12. Carre 1985; the title of her lecture (with M.T. Cipriano) at the conference in Siena (22-24 May, 1986) was Production et typologie des amphores de la côte adriatique de l'Italie.

13. The drawings are the work of Hajnal Gödér, Ferenc Gelencsér and Ágnes Vári; the photography is the work of András Dabasi and Zsuzsa Erdökürti.

14. Plesničar-Gec 1972: grave 129., Table 34 No. 1., cf. Tchernia 1971.

15. The Salla finds were dated by Ferenc Redö.

16. Magdolna Medgyes, personal communication.

17. Medgyes 1977, 180. Table 8 No. 2.

18. Two amphorae (from Aquileia and Ostia) were also analysed. Their components closely resemble those of amphorae along the Amber Route.

19. Baldacci 1967. 15.

20. Buchi 1973, 547-553.

21. It may be supposed that some base sherds of the same type are present at Emona; no. 311 from Savaria is listed as Portorecanatian but it may belong to this type as well.

22. Tassaux 1982, 244.

23. PIR[1] 259; PIR[2] 5.1. 7-8. no. 30, PWRE, 12.1, 1924, 396.

24. PIR[2] 5.1.8. no. 31.

25. CIL. 5, 698.; Tassaux 1982, 245; Buchi 1974, n. 42.

26. Gnirs 1910A, 95; Gnirs 1910B, 79-80.

27. Gnirs 1910A, 99-

28. Gnirs 1904, 145-146.

29. Tassaux 1982, 254.

30. Tassaux 1982, 255-256.

31. The amphorae from the Magdalensberg are described by Fuchs 1978.

    ADEL(phus) Catalogue 960; BAR(...) Cat, 961-962; BARB(...) Cat, 963; BARD(...) Cat, 965; COM(...) Cat, 966-967; EVHAR(istus) Cat, 969-970; FELIX Cat, 971-975; FELIX.PET Cat, 976-982; FELIX.SCR Cat, 983-984; FELIX.SER Cat, 985; FEL.TVR Cat, 986-987; HER(...) Cat, 989-990; SPERATVS Cat, 992-995; VRBA(nus) Cat, 996-1000; VIAT(or) Cat, 1057.

32. Fuchs 1978, Band 1, 55; Raum OR, 40.

33. Lamboglia 1955, 245.

34. Abramič 1925, 110.

35. Baldacci 1967, 34.

36. Baldacci 1967, 31; Pais 1077, 4.

37. Pais 1077. 11. AEk, HER; 20. ADEL; 66. EVHAR; 70a FEX.PE; 78. HER; 87. IAEk, NIA; 88a C.LAEk.BASS; 88b C.LAEk, FELIX; 88c CLAEk. BA , FEIX.SCR; 90. LAE, HOM; 93. C LAEk.BAS, SPERATVS; 137. SPER; 152. VIAT.

38. Nagy 1948, 108.

39. Brukner 1981, Table 159 no. 47. Teutoburgium; no. 36. Aquincum.

40. Tassaux 1982, 245.

41. Degrassi 1942, 17-22.

42. Plinius NH, 36.5.

43. Tassaux 1982, 260.

44. Callender 1965, no. 365. 104.

45. Baldacci 1967, 35-36; Tassaux 1982, 255-259.

46. Tacitus Hist, 1.73.

47. Tassaux 1983, 203-204.

48. Abramič 1925, 110.

49. Callender 1965, 114 no. 463b.

50. Baldacci 1967, 33.

51. Tassaux 1983, 203.

52. Ulbert 1959, 52 Abb.9.3.

53. Schönberger 1978, 274 no. 533, Abb.82.1.

54. Schallmayer 1983, 295-299.

55. Pais 1077, 16. 51a-b.

56. Szőnyi 1981, 98 picture 8. The reading of the stamp is CAL.CRISPINILLAE

57. Inventory no. 447.

58. Fuchs 1978, Cat. 890-892.

59. Tassaux 1982, 204.

60. Gregorutti 1886, 230.

61. Inventory no. 302.

62. Baldacci 1967, 49.

63. Gnirs 1910, on the left of picture 26.

64. Bezeczky 1984, 158, Pl. 4.

65. The numbers refer to the weights of various foodstuffs; 100 roman pounds = 32.745kg; LXI = 19.97kg; LXXII = 23.9kg.

66. Zevi 1966, 217-219.

67. Baldacci 1967; Buchi 1973.

68. Alzinger 1955, Cat. 145.

69. On Dressel 6B cf. Fuchs 1978, Cat. 1099.

70. Hesnard 1980, 144.

71. Zevi 1966, 230.

72. Zevi 1966, 231.

73. Jobst & Piccottini 1986, 272-273 no. 15.

74. Zevi 1966, 233.

75. Pliny NH, 31 93.4; Martial 13.102.

76. CIL 3. 9387-9406.

77. CIL 15.2. 4686-4695; 4697-4799; 4701; 4705-4706; 4709; 4718.

78. CIL 4. 5657-5659; 5663-5679.

79. Bohn 1926, 202-203.

80. Bohn 1926, 203.4; Columella 12.6.

81. Barkóczi & Bónis 1954, 167. 86, Table 24. no. 11. inventory no. 9.1951.34.; rim, inv.no. 2.1951.145; handle fr. 2.1951.122; all these amphorae are in the Hungarian National Museum at Budapest.

82. The excavation of László Kocsis, handle fr.

83. Tchernia 1967, 224; Zevi 1966, 221.

84. Gábler & Kelemen 1984, and some other handle and rim sherds in Gorsium, inv.nos. 78.145/835.5; 60.270.10; 78.250/805.4; 79.150/805.4; unpublished.

85. Tchernia 1967, 224.

86. Pleničar-Gec 1972, grave 969. no.1.

87. Abramič 1925, 110.

88. Carnuntum Jahrbuch 1957, 68; 1959, 87; 1960, 87.

89. Bonsor 1931, T.XXXV. no. 160.

90. Callender 1965, 106 no. 392.

91. Bonsor 1931, T.XXXIX. no. 334; Callender 1965, 228 no. 1473.

92. The Amber Route finds demonstrate the continuity of the oil trade, so one may doubt if it only began under the Severi, cf. Gábler & Kelemen 1984, 136-137.

93. Pélichet 1946, 192.

94. inv.nos. 72.160/750.3; 72.210/830.5; unpublished.

95. inv.nos. 2.1951.146; unpublished.

96. The excavation of László Kocsis, rim and handle fr.

97. Grace 1961, fig. 62.

98. Peacock 1977, 266-270.

99. Baldacci 1969, 114.

100. Peacock, 1977, 270.

101. Schörgendorfer 1942, 65-66.

102. Baldacci 1972, 22-23, 27; a sherd from Aquileia, cf. Carre 1985, 231 fig. 6.

103. From Aquincum, excavated by László Kocsis.

104. Mikl 1960, 156, T.II.12.

105. Neumann 1967, 46. T.XXIX. no. 2.

106. Schörgendorfer 1942, 65-66.

107. Alzinger 1955, 18 no. 143.

108. Baldacci 1972, 28.

109. M. Medgyes, personal communication.

110. cf. note 103.

111. inv.nos. 69.190/825.3; 82.165/865.3.

112. Baldacci 1972, 27-28.

113. Polaschek PWRE, 1, 1936, col. 1040.

114. Alzinger 1955, 18. nos. 140-142.

115. Bojović 1977, 58. no. 559, T.LXIV.

116. Bojović 1977, 58.

117. Degrassi 1953, 60; *idem* 1956, 108-109.

118. Baldacci 1972, 28.

119. Martial 12. 43; Pliny NH, 15.8.

120. Degrassi 1956, 107.

121. Tchernia 1967, 228.

122. Degrassi 1953, 57; *idem* 1956, 107; Zevi 1967, 30.

123. Boltin Tome 1974; Bezeczky 1985; Barkóczi & Bónis 1954, 177. Abb.28:22, Taf. LVIII.14.

124. Carre 1985, 232-234. figs. 7-8.

125. Bohn 1926, no. 321.

126. Excavations by Tibor Nagy, inv.no. 71.3.471, unpublished.

127. Ettlinger 1977, fig. 1.

128. Fuchs 1978, Cat. 828.

129. Mercando 1974, figs. 66, 79, 116, 121, 133, 158; *idem* 1979, figs. 138, 139, 176, 177.

130. Schönberger 1978, no. 523. T.103.

131. Vrsalović 1974, no. 113.

132. Mercando 1979, fig. 176, stamp: LCSREV

133. Mercando 1979, 266, n. 340.

134. C. Panella, personal communication; cf. Keay 1984, XLII.

135. Grace 1961, 67.

136. Riley 1982, 226.

137. Bónis 1942, 232.

138. Panella 1974, 596-597; Oelmann 1914, 65. fig. 47; Käpitan 1972, 246, 248.

139. Riley 1982, 192.

140. Baldacci 1967, 14.

141. Strab 4. 5.10.

142. Lamboglia 1955, 264.

143. Guenoche & Hesnard 1983, 191-193.

144. Zevi 1966, 216.

145. Cassiodorus Variae 12.22.1.

146. cf. Józsa & Szakmány: Dressel 2-4 amphorae nos. 7 and 31.

147. L. Plesničar-Gec, personal communication.

148. Mócsy 1980, 79-80.

149. Póczy 1955, 16.

150. We studied the finds at the villa in 1984.

151. In view of this, the production of wine must have been less significant in Istria than that of olive oil.

152. Bónis 1969, Abb. 17. no. 15.

153. Cassius Dio 49. 36. Boiss.

154. Pais 1077, 10, 14,

155. Nos. 60, 63, 66-67.

156. Fuchs 1978, Cat, 913-923.

157. CIL 3. 12010.12.

158. No. 68.

159. Fuchs 1978, Cat. 899-901.

Supplement 1

Petrology

S. Józsa & Gy. Szakmány

We have undertaken mineralogical and petrographical investigations on amphorae collected along the Amber Route for several years. The results of the analyses, classification of the fragments and the materials applied by the different workshops are described below.

Qualitative and quantitative analyses were made on 30 μm thick thin-sections with a polarizing microscope. Matrix, clasts and limonite were separated and their ratios determined. Clasts are grains larger than 10 μm in diameter and therefore recognizable under the microscope. When the mineral and rock grains were identified, their percentages were calculated. We applied a square grid for the quantitative analyses. The average dimensions of the clasts were also determined.

The following data was used for interpretation:

- quantity of quartz;
- presence and quantity of K-feldspars (orthoclase, sanidine, microcline);
- presence and quantity of plagioclase;
- presence and quantity of carbonates (sparite, micrite, calcite fossils);
- eventual presence and quantity of other rock fragments (volcanic or metamorphic rocks);
- presence and quantity of other components (augite, hornblende, chlorite, opaque minerals);
- ratio of clasts compared to the matric and limonite.

Micas (biotite and muscovite) were not considered for the interpretation, since these minerals occur in almost all amphora types. Differentiation of biotite and limonite is highly problematic, since biotite is altered to limonite during firing of the amphorae. Quantities of biotite and muscovite were more or less the same in all samples.

Accessory minerals (zircon, tourmaline, zoisite, garnet, apatite) occur in very small quantities only (less than 1-2%). Their appearance in thin-section is unsatisfactory and such components cannot therefore contribute to the results.

Limonite may occur because of the alteration of several silicate minerals (biotite, pyroxenes, hornblende, etc.). It forms larger grains because of the uneven distribution of iron in the clay, or else indistinct inclusions. The quality of limonite was not considered in the interpretation because of its irregular distribution throughout the material. The colour of the matric can be different, depending on the quality of the clay (iron content and source) and on the firing temperature.

Comparing the fragments, we can say that in some cases mineralogical and petrological investigations alone do not give adequate enough evidence to differentiate the groups. The restricted area of thin-sections, poor preservation of minerals and final appearance of the characteristic and special mineral components, are the main problems in characterizing the sherds. Presence or absence of certain components is not always decisive in the recognition of a workshop or group of workshops, since the composition of the clay may vary within a quarry and one workshop could have used clay from several sources.

Petrology of the fragments by types and workshops

Dressel 2-4

These amphorae have a high clast content (3.1 - 21.8%). Nearly all samples contain plagioclase, rarely in large quantities (15.3%). Quartz is more common (10.9 - 70.1%), calcitic fossils are almost absent (except in two amphorae).

I. This group is characterized by few clasts (3.1 - 4.3%), much limonite (2.1 - 3.3%) and orthoclase (5.5 - 16.7%), and little carbonate (0 - 2%).

No. 20 (Salla 83.5.106).

No. 19 (Salla 83.5.135).

No. 10 (Poetovio 3685).

The Salla piece (No. 20) contains a high proportion of calcitic fossils (28.2%).

II. Moderate quantities of quartz (38 - 55%), much plagioclase (15 - 24%) and fewer carbonates (from trace amounts up to 11%). Chlorite and chloritephyllite are also present.

No. 16 (Salla 83.5.104).

No. 21 (Salla 83.5.53).

No. 31 (Savaria 77.2.1811).

Amphora No. 12 (Poetovio R 500) could have been included in this group on the basis of its chlorite content. But its carbonate content is very high; the plagioclasse content is sparse (3.2%).

III. The highest amount of clasts was found in this group (15 - 22%). Significant amounts of quartzite (44 - 53%) and sparse quartz (11 - 28%) are present. Carbonate content: 7.7-17.6%. Both amphorae contain opaline fossils.

No. 13 (Salla 83.5.55).

No. 14 (Salla 83.5.56).

IV. Sparse clasts (4.6 - 7.3%); abundant quartz content (60 - 63%); sparse amounts of K-feldspar (3.2 - 4.8%) and moderate carbonates (micrite only: 11 - 17%).

No. 7 (Poetovio).

No. 30 (Savaria 77.2.5941).

						Dressel 2-4										Dressel 5	
	Salla No. 21	Salla No. 13	Salla No. 14	Salla No. 16	Salla No. 17	Salla No. 20	Salla No. 19	Savaria No. 26	Savaria No. 32	Savaria No. 31	Savaria No. 30	Ad Arrabonem No. 22	Poetovio No. 10	Poetovio No. 7	Poetovio No. 12	Savaria No. 45	Poetovio No. 44
Matrix	87,8	78,1	84,6	88,6	89,0	93,6	93,9	86,3	87,5	85,5	92,0	9o,7	92,5	95,0	90,4	88,3	91,5
Limonite	0,7	0,1	0,1	0,8	0,7	2,1	3,0	0,2	2,0	0,7	0,7	1,2	3,3	0,4	1,1	1,0	0,9
Cast	11,5	21,8	15,3	10,6	10,3	4,3	3,1	13,5	10,5	13,8	7,3	8,1	4,2	4,6	8,5	10,7	7,6
Distribution of clasts in %																	
Quartz	38,7	10,9	27,7	55,8	34,6	30,6	35,6	70,1	57,0	39,8	60,3	62,4	49,0	63,3	30,9	12,4	12,0
Orthoclase	-	1,1	4,4	-	-	16,7	5,5	5,0	13,3	-	-	-	9,1	4,8	5,5	-	1,0
Plagioclase	15,3	4,7	tr	18,6	1,6	3,7	tr	6,6	6,1	24,4	3,2	2,9	6,9	4,8	3,2	tr	tr
Sanidine	-	-	-	-	53,6	-	-	-	-	2,8	3,1	3,0	-	-	1,2	-	-
Muscovite	13,0	tr	3,0	3,2	4,1	-	1,0	6,5	3,4	1,9	4,9	6,0	16,1	1,8	8,6	tr	1,0
Biotite	5,8	1,2	2,6	4,8	4,7	15,0	30,5	10,8	12,0	3,1	10,1	9,0	18,9	9,6	13,0	7,8	1,9
Pale biotite	-	-	-	-	-	-	-	-	-	-	-	14,5	-	-	-	-	-
Chlorite	-	-	-	tr	-	-	-	-	-	-	-	-	-	-	2,1	-	-
Zoisite	1,4	tr	tr	-	-	tr	-	tr	tr	-	tr	tr	tr	-	1,1	-	-
Tourmaline	tr	-	-	-	-	-	-	tr	tr	-	-	-	tr	-	-	-	-
Hypersthene	-	-	-	-	-	-	-	-	-	tr	-	-	-	-	-	-	-
Apatite	-	-	tr	-	tr	-	-	-	-	-	tr	-	-	-	-	-	-
Zircon	-	-	-	-	tr	tr	-	-	tr	tr	tr	-	-	-	-	-	-
Garnet	-	-	-	1,1	-	-	-	1,0	2,6	-	-	tr	-	-	-	-	-
Titanite	tr	tr	-	-	-	tr	-	tr	-	tr	-	-	tr	-	-	-	-
Augite	-	-	-	-	-	-	1,4	-	-	-	-	-	-	-	-	1,9	2,1
Opaque constituent	-	tr	-	1,3	tr	-	-	-	-	3,7	-	-	-	-	-	tr	-
Epidote	-	-	-	-	-	-	-	-	-	-	-	-	-	tr	-	-	-
Quartzite	5,0	53,0	43,9	15,2	1,4	5,8	tr	tr	5,6	14,4	1,4	tr	tr	-	1,0	2,8	2,3
Sparite	7,6	-	-	tr	-	-	2,1	-	-	1,8	-	-	-	-	33,4	-	-
Micrite	3,4	7,7	17,6	-	-	-	tr	-	-	2,1	17,0	-	-	11,0	-	59,8	79,7
Andesite	-	-	-	-	-	-	-	-	-	-	-	-	-	-	-	6,9	-
Shale	-	-	-	-	-	tr	-	-	-	-	-	2,2	-	-	-	8,4	-
Clay granulate	-	20,2	-	-	-	-	-	-	-	-	-	-	-	-	-	-	-
Graphic granite	-	-	-	-	-	-	-	-	-	-	-	tr	-	-	-	-	-
Mica-schist	1,4	-	-	-	-	-	-	-	-	3,7	-	-	-	4,7	-	-	-
Chlorite-phyllite	8,4	-	-	-	-	-	8,2	-	-	2,3	-	-	-	-	-	-	-
Calcitic fossil	-	-	-	-	-	28,2	-	-	-	-	-	-	-	-	-	-	-
Opalic fossil	-	1,2	o,8	-	tr	-	-	-	-	-	-	-	-	-	-	-	-
Quartzite fossil	-	-	-	-	-	-	15,7	tr	-	-	-	-	-	-	-	-	-

Table 1 Dressel 2–5 amphorae

	Dressel 2-4															Dressel 5	
Measure of clasts /µm/	Salla No. 21	Salla No. 13	Salla No. 14	Salla No. 16	Salla No. 17	Salla No. 20	Salla No. 19	Savaria No. 26	Savaria No. 32	Savaria No. 31	Savaria No. 30	Ad Arrabonem No. 22	Poetovio No. 10	Poetovio No. 7	Poetovio No. 12	Savaria No. 45	Poetovio No. 44
Quartz	80-250	50-200	100-500	60-200	40-280	40-100	30-50	60-200	80-250	150-500	100-500	60-300	20-40	50-100	40-80	60-120	40-80
Orthoclase	-	40-80	80-350	-	-	30-60	20-40	80-200	150-200	-	-	-	20-40	40-60	50-150	-	60-100
Plagioclase	80-300	50-200	50-300	60-300	30-100	100-150	10-20	60-200	80-200	150-350	100-200	50-100	100-150	60-110	40-60	40-70	100-200
Sanidine	-	-	-	-	40-300	-	-	-	-	100-300	100-200	60-250	-	-	50-80	-	-
Muscovite	60-150	40-60	30-80	50-150	30-80	50-150	30-80	-	30-60	40-120	60-150	40-150	40-120	20-50	60-250	40-60	40-80
Biotite	50-150	40-60	40-80	60-250	20-60	60-200	40-80	40-100	40-80	80-300	60-250	50-150	40-120	10-30	50-120	60-80	60-140
Pale biotite	-	-	-	-	-	-	-	-	-	-	-	40-60	-	-	-	-	-
Chlorite	-	-	-	40-100	-	-	-	-	-	-	-	-	-	-	30-60	-	-
Zoisite	20-60	10-20	20-30	-	-	~20	-	40-60	10-30	-	20-30	30-40	20-40	-	20-40	-	-
Tourmaline	20-30	-	-	-	-	-	-	20-40	20-30	-	-	-	15-30	-	-	-	-
Hypersthene	-	-	-	-	-	-	-	-	-	120-250	-	-	-	-	-	-	-
Apatite	-	-	~60	-	20-30	-	-	-	-	-	15-25	-	-	-	-	-	-
Zircon	-	-	-	-	10-20	10-20	-	-	10-20	10-20	10-20	-	-	-	-	-	-
Garnet	-	-	-	50-80	-	-	-	50-60	150-250	-	-	100-120	-	-	-	-	-
Titanite	20-30	10-30	-	-	-	10-15	-	20-40	-	20-30	-	-	20-30	-	-	-	-
Augite	-	-	-	-	-	-	40-60	-	-	-	-	-	-	-	-	60-100	30-50
Opaque constituent	-	40-60	-	40-80	40-50	-	-	-	-	60-150	-	-	-	-	-	20-40	-
Epidote	-	-	-	-	-	-	-	-	-	-	-	-	-	20-30	-	-	-
Quartzite	100-250	50-450	100-450	100-350	50-80	40-80	30-60	40-60	60-150	100-400	40-80	100-150	80-120	-	80-120	60-140	100-300
Sparite	80-250	-	-	20-30	-	-	60-80	-	-	100-250	-	-	-	-	40-80	-	-
Micrite	60-350	50-150	100-300	-	-	-	100-500	-	-	100-450	150-600	-	-	10-30	-	100-200	200-400
Andesite	-	-	-	-	-	-	-	-	-	-	-	-	-	-	-	100-200	-
Shale	-	-	-	-	-	80-90	-	-	-	-	-	100-250	-	-	-	60-100	-
Clay granulate	-	100-300	-	-	-	-	-	-	-	-	-	-	-	-	-	-	-
Graphic granite	-	-	-	-	-	-	-	-	-	-	-	40-60	-	-	-	-	-
Mica-schist	100-150	-	-	-	-	-	-	-	-	100-350	-	-	-	40-80	-	-	-
Chlorite-phyllite	150-250	-	-	-	-	-	40-80	-	-	150-300	-	-	-	-	-	-	-
Calcitic fossil	-	-	-	-	-	80-250	-	-	-	-	-	-	-	-	-	-	-
Opalic fossil	-	60-90	80-200	-	40-60	-	-	-	-	-	-	-	-	-	-	-	-
Quartzite fossil	-	-	-	-	-	-	20-60	40-50	-	-	-	-	-	-	-	-	-

Table 2 Dressel 2-5 amphorae

V. Moderate clasts (8.1 - 16.4%), abundant quartz (57 - 70%) and no carbonates. Plagioclase and K-feldspar are present, the former in lesser (2.9 - 6.6%), the latter in greater amounts (5 - 13.3%).

No. 22 (Ad Arrabonem).

No. 26 (Savaria 74.4.1370).

No. 32 (Savaria 76.2.9).

VI. Extremely high sanidine content.

No. 17 (Salla 83.5.105).

Dressel 5

This type is characterized by small amounts of quartz (7.7 - 12.4%) and sparse augite (1.9 - 6.3%) with much micrite (59.8 - 79.7%).

No. 44 (Poetovio).

No. 45 (Savaria 69.6.6.).

Dressel 6B

Workshop of C. Laecanius Bassus

Four samples were taken from the rim, neck, body and handle of an amphora from this workshop during restoration. (This amphora comes not from the Amber Route but from Aquincum; inventory no. 56.63.21; stamps: C.LAEK.B; BAR). There are no carbonates in the handle. K-feldspars are present in the body and handle only, plagioclases in the rim and neck only. Accessory minerals also differ between body and handle, and the rim and neck, although such differences are very small. The most significant feature is the present or absence of carbonates. This case showed us the importance of the location of the sample on the amphora.

Amphorae from this workshop contain sparse to medium amounts of clasts (3 - 9.7%), very different amounts of quartz (23.6 - 87.6%) and plagioclases in trace amounts, or up to 1 - 2%. Almost all samples contain carbonates (sparite, micrite and calcitic fossils), rarely in significant amounts (32.4%). Calcitic fossils are present in far more samples than in the Dressel 2-4 amphorae. The quantities of quartzite are modest, with rare exceptions.

	Rim	Neck	Body	Handle	Rim	Neck	Body	Handle
Matrix	92,1	94,3	94,0	96,1				
Limonite	1,1	1,3	0,7	0,2				
Clast	6,8	4,4	5,3	3,7				
Distribution of clasts in %					Measure of clasts /μm/			
Quartz	32,1	52,7	41,7	60,6	50-80	40-160	40-80	20-120
Orthoclase	-	-	1,4	3,2	-	-	50-100	20-50
Plagioclase	tr	1,4	-	-	30-40	40-60	-	-
Muscovite	1,0	6,8	3,3	11,1	30-50	40-80	40-60	20-150
Biotite	6,2	13,9	11,8	24,0	60-100	40-80	40-100	10-80
Chlorite	1,9	-	-	-	30-40	-	-	-
Zoisite	1,7	1,1	-	-	30-60	30-40	-	-
Tourmaline	tr	-	tr	tr	10-20	-	20-30	20-30
Apatite	-	-	tr	-	-	-	20-30	-
Zircon	-	tr	tr	tr	-	10-20	10-20	10-20
Titanite	-	-	-	tr	-	-	-	20-40
Opaque constituent	-	-	tr	tr	-	-	10-20	10-20
Epidote	-	-	tr	-	-	-	20-30	-
Quarzite	4,9	3,1	11,4	1,1	50-100	40-80	60-80	60-80
Sparite	13,8	3,4	-	-	40-60	40-80	-	-
Micrite	-	-	30,4	-	-	-	100-200	-
Calcitic fossil	38,4	17,6	tr	-	250-500	150-400	300-500	

Table 3 C.LAEK.B; BAR stamped amphora from Aquincum

I. Abundant quartz (71.4 - 87.6%), sparse carbonate (traces - 3.9%), relatively little quartzite (traces - 8.4%). Calcitic fossils are absent, except in No. 75.

No. 75 FELC (Poetovio 3671).

HER (Aquileia 430).

No. 51 SPERATVS (Emona G3223).

SPERATVS (Aquileia 395).

No. 150 VIAT (Savaria 65.52.1).

VIAT (Aquileia 389).

II. Abundant quartz (50.4 - 63.8%); relatively large amounts of orthoclase (2.6 - 9.4%). Medium carbonate content (2.5 - 23%), few calcitic fossils (traces to 1%). Medium quartzite content (2.5 - 6.1%).

No. 76 CLYMEN (Poetovio 10094).

FELIX.SER (Aquileia 376).

HOM (Aquileia 373).

VIAT (Aquileia 378).

VIAT (Aquileia 443).

III. A relatively high amount of clasts (6.9 - 9%), quartz is sparse to common (26 - 50.1%), but abundant opaque minerals (12.6 - 21.5%). Few plagioclases (1.1 - 4.3%), carbonates are common to abundant (4.1 - 31.8%) with sparse quantities of quartzite (traces - 1.1%).

HER (Aquileia).

HER (Aquileia 379).

LAEK (Aquileia 375).

IV. Quartz is sparse to common (23.6 - 32.1%), carbonate is common (6 - 64.5%), sometimes with abundant calcitic fossils (1 - 38.4%). Plagioclases are sparse (less than 1.3%).

No. 54	FI/...	(Emona).
	BAR	(Aquincum 56.63.21).
	CRESCENTIS	(Aquileia 391).
	FELIX.SER	(Aquileia 371).
	FEX.PE	(Aquileia).
	PIERI	(Aquileia 377).

V. Quartz is common (42 - 53%), carbonates are significant (18 - 50%), plagioclase is sparse (less than 1.7%) with little quartzite (2.3 - 8.5%).

	ADEL	(Aquileia 399).
	FELIX.SER	(Aquileia 401).
No. 97	FELIX.PE	(Salla 84.1.2).
	SPERATVS	(Aquileia 374).
No. 52	VIAT	(Emona F 249).
	LAEKB	(Aquileia 449).
	LAI	(Aquileia 442).
	LAEK L	(Aquileia 6177).

VI. Abundant quartz (62.2 - 70.2%), medium carbonate (10.3%) and plagioclase content (1.5 - 2%) with relatively common quartzite (4.2 - 9.9%).

	BAR	(Aquileia 433).
	CLAEK	(Aquileia 452).

VII. Quartz is common (47.8 - 57.8%), so too is quartzite (8.9 - 12.8%), carbonate is sparse (1 - 5.1%) with a relatively high biotite content (16%). Plagioclase is present in only one sample.

	LAEK	(Aquileia 400).
No. 74	LAEKB	(Poetovio 3435).

	ADEL Aquileia 399	BAR Aquincum 56.63.21.	BAR Aquileia 433	CLYMEN Poetovio No. 76	CRESCENTIS Aquileia 391	FELIX.SER Aquileia 401	FELIX.SER Aquileia 376	FELIX.SER Aquileia 371	FELIX.PE Aquileia 441	FELIX.PE Salla No. 97	FELIX Emona No. 54	FELC Poetovio No. 75	HER Aquileia	HER Aquileia 379	HER Aquileia 430	HOM Aquileia 373	PIERI Aquileia 377
Matrix	91,5	92,1	94,2	94,1	92,8	94,0	93,1	88,5	92,1	94,9	89,0	96,0	91,9	92,9	92,0	92,5	93,1
Limonite	0,9	1,1	1,4	1,4	0,1	1,3	0,3	1,8	0,5	0,4	2,6	1,0	-	0,2	0,5	0,4	0,7
Clast	7,6	6,8	4,4	4,5	7,1	4,7	6,6	9,7	7,4	4,7	8,4	3,0	8,1	6,9	7,5	7,1	6,2
Distributin of clasts in %																	
Quartz	53,0	32,1	70,2	53,4	23,6	44,6	63,8	37,2	25,7	42,0	26,0	74,9	50,1	26,0	71,4	50,4	25,9
Orthoclase	2,8	-	3,9	6,3	-	-	2,6	1,0	1,0	tr	-	1,3	tr	-	2,8	3,6	2,5
Plagioclase	1,7	tr	1,5	-	-	-	-	1,3	-	tr	-	tr	1,0	1,1	3,9	tr	1,3
Muscovite	3,0	1,0	5,6	5,4	7,6	4,9	3,8	3,0	4,1	8,7	7,5	7,7	2,0	1,9	3,6	6,0	1,9
Biotite	4,1	6,2	4,4	7,4	2,2	5,7	10,5	4,2	7,2	19,1	6,0	3,2	7,3	3,7	8,9	10,0	1,1
Chlorite	-	1,9	-	-	-	-	-	-	-	-	-	-	-	-	-	-	-
Zoisite-clinozoisite	tr	1,7	tr	2,0	-	tr	-	tr	tr	-	tr	-	2,0	-	tr	tr	-
Tourmaline	-	tr	-	-	-	1,6	1,7	tr	-	-	-	-	tr	-	tr	-	-
Apatite	-	-	-	-	-	-	-	-	-	tr	-	-	tr	-	-	-	-
Zircon	tr	-	-	tr	-	-	-	tr	-	tr	tr	-	tr	-	-	-	tr
Garnet	tr	-	tr	tr	tr	-	-	-	tr	tr	tr	-	-	-	-	tr	-
Titanite	tr	-	-	-	-	-	tr	-	-	-	-	-	1,0	-	-	tr	-
Opaque constituent	-	-	-	-	-	-	-	-	tr	tr	-	-	12,6	21,5	5,0	-	-
Epidote	-	-	-	-	-	tr	-	-	-	-	-	-	-	-	tr	-	-
Quartzite	-	4,9	4,2	-	-	2,9	3,9	2,5	3,5	8,2	1,2	3,2	-	tr	1,0	4,5	tr
Sparite	32,4	13,8	-	21,2	1,1	27,3	7,5	32,3	31,7	16,2	8,4	1,9	tr	31,8	-	2,5	52,5
Micrite	3,0	-	10,2	2,6	64,5	11,9	4,8	9,1	11,9	5,8	28,8	2,6	24,0	3,9	1,7	23,0	6,8
Phyllite	-	-	-	-	-	-	-	-	-	-	-	1,3	-	-	-	-	-
Chlorite-phyllite	-	-	-	-	-	-	-	-	-	-	-	-	-	-	1,7	-	-
Clay granulate	-	-	-	-	-	-	tr	-	-	-	-	-	-	-	-	-	-
Calcitic fossil	-	38,4	-	1,7	1,0	1,1	1,4	7,3	14,9	-	22,1	3,9	-	10,1	-	tr	8,0
Opalic fossil	tr	-	-	tr	-	-	-	2,1	-	-	-	-	tr	-	-	tr	-
Quartzite fossil	-	-	-	-	-	-	-	-	-	-	-	-	-	-	-	tr	-

Table 4 Laecanius workshops amphorae

Measure of clasts /µm/	ADEL Aquileia 399	BAR Aquincum 56.63.21.	BAR Aquileia 433	CLYMEN Poetovio No. 76	CRESCENTIS Aquileia 391	FELIX.SER Aquileia 401	FELIX.SER Aquileia 376	FELIX.SER Aquileia 371	FELIX.PE Aquileia 441	FELIX.PE Salla No. 97	FELIX Emona No. 54	FELC Poetovio No. 75	HER Aquileia	HER Aquileia 379	HER Aquileia 430	HOM Aquileia 373	PIERI Aquileia 377
Quartz	60-150	50-80	40-100	40-80	20-30	40-80	20-100	40-100	30-60	30-80	20-80	40-80	50-130	40-80	10-50	20-50	40-80
Orthoclase	80-250	-	40-60	40-80	-	-	40-80	60-120	20-50	30-40	-	40-60	30-50	-	20-50	20-60	40-120
Plagioclase	60-100	30-40	40-80	-	-	-	-	40-100	-	20-50	-	20-50	20-40	40-60	20-40	20-30	60-120
Muscovite	60-100	30-50	40-80	60-100	20-50	40-80	60-70	40-80	20-40	10-50	50-80	30-70	20-40	40-60	20-40	40-70	40-80
Biotite	40-80	60-100	40-60	40-80	80-140	40-80	40-50	40-60	30-50	10-100	50-80	20-40	50-70	40-60	20-40	40-60	40-80
Chlorite	-	30-40	-	-	-	-	-	-	-	-	-	-	-	-	-	-	-
Zoisite-clinozoisite	20-40	30-60	20-50	20-30	-	20-40	-	10-40	20-40	-	15-20	-	60-80	-	20-40	20-30	-
Tourmaline	-	10-20	-	-	-	30-50	~30	30-40	-	-	-	-	30-40	-	20-30	-	-
Apatite	-	-	-	-	-	-	-	-	-	~20	-	-	20-30	-	-	-	-
Zircon	15-30	-	-	10-20	-	-	-	10-20	-	10-20	20-30	-	10-20	-	-	-	10-20
Garnet	20-50	-	20-40	10-20	20-30	-	-	-	20-40	20-40	60-70	-	-	-	-	20-50	-
Titanite	30-60	-	-	-	-	-	20-30	-	-	-	-	-	60-80	-	-	30-50	-
Opaque constituent	-	-	-	-	-	-	-	-	30-70	20-30	-	-	50-110	80-200	20-40	-	-
Epidote	-	-	-	-	-	20-30	-	-	-	-	-	-	-	-	10-30	-	-
Quartzite	-	50-100	40-80	-	-	40-60	40-60	40-80	40-80	40-60	30-50	40-60	-	60-80	40-100	30-60	40-80
Sparite	60-300	40-60	-	40-80	60-80	40-60	100-150	40-150	40-80	20-60	30-100	30-50	100-130	60-150	-	40-80	100-200
Micrite	60-150	-	80-150	40-60	100-200	40-120	50-150	40-150	300-600	20-150	100-300	20-50	200-300	80-300	30-60	50-80	60-100
Phyllite	-	-	-	-	-	-	-	-	-	-	-	~50	-	-	-	-	-
Chlorite-phyllite	-	-	-	-	-	-	-	-	-	-	-	-	-	-	30-70	-	-
Clay granulate	-	-	-	-	-	-	100-200	-	-	-	-	-	-	-	-	-	-
Calcitic fossil	-	250-500	-	60-100	200-500	40-60	10-40	60-200	120-220	-	100-200	50-80	-	100-400	-	60-180	80-150
Opalic fossil	30-80	-	-	20-30	-	-	-	40-50	-	-	-	-	~50	-	-	40-60	-
Quartzite fossil	-	-	-	-	-	-	-	-	-	-	-	-	-	-	-	20-40	-

Table 4a Laecanius workshops amphorae

	SPERATVS Emona No. 51	SPERATVS Aquileia 395	SPERATVS Aquileia 374	VIAT Aquileia 378	VIAT Savaria No. 150	VIAT Emona No. 52	VIAT Aquileia 389	VIAT Aquileia 443	CLAEK Aquileia 452	LAEKB Aquileia 449	LAEK Aquileia 400	LAEKB Poetovio No. 74	LAE Salla No. 96	LAEK Aquileia 375	LAI Aquileia 442	LAEK L Aquileia 6177
Matrix	93,4	95,5	95,1	95,8	91,7	93,3	96,1	93,2	92,0	91,6	94,8	96,9	89,7	90,0	91,7	92,8
Limonite	1,7	1,2	1,5	0,1	1,5	0,9	0,1	1,1	1,8	0,3	0,4	0,2	1,3	1,0	1,1	0,3
Clast	4,9	3,3	3,4	4,1	6,8	5,8	3,8	5,7	6,2	8,1	4,8	2,9	9,0	9,0	7,2	6,9
Distribution of clasts in %																
Quartz	87,6	82,2	50,7	52,2	75,7	46,7	72,8	56,3	62,2	46,7	47,8	57,8	50,0	42,3	42,5	47,9
Orthoclase	-	7,4	-	9,4	6,9	1,7	2,1	6,6	-	-	6,8	1,8	3,0	-	-	1,1
Plagioclase	-	tr	1,2	tr	tr	-	tr	tr	2,0	tr	-	tr	tr	4,3	tr	1,4
Microcline	-	-	-	-	-	-	-	-	-	-	-	-	-	tr	-	-
Muscovite	5,0	3,8	7,8	7,0	9,5	1,0	8,9	6,6	4,1	17,9	10,6	9,2	5,0	2,7	3,1	4,2
Biotite	2,5	5,6	8,9	14,4	5,9	5,1	7,8	5,5	4,9	9,9	15,7	15,6	16,0	8,6	2,9	7,6
Chlorite	-	-	-	-	-	-	-	-	-	-	-	-	-	4,6	-	-
Zoisite-clinozoisite	1,7	-	1,0	1,5	1,0	tr	tr	tr	tr	tr	-	tr	1,0	tr	-	1,0
Tourmaline	-	-	1,0	1,0	tr	-	-	tr	-	-	-	tr	tr	-	tr	1,0
Rutile	-	-	-	-	-	tr	-	-	-	-	-	-	-	-	-	-
Apatite	tr	-	-	-	-	-	-	-	-	-	-	-	-	-	-	-
Lazurite	-	-	-	-	-	-	-	-	-	-	-	tr	-	-	-	-
Zircon	1,6	-	tr	tr	tr	tr	-	-	tr	tr	-	tr	tr	tr	-	-
Garnet	-	-	-	-	tr	-	tr	-	tr	-	-	-	tr	-	-	-
Titanite	-	-	-	-	-	-	tr	tr	tr	-	-	-	-	-	-	-
Opaque constituent	-	-	-	-	-	-	-	-	-	1,2	-	-	-	20,0	-	-
Epidote	-	-	-	-	tr	-	-	1,6	-	tr	-	tr	-	-	-	-
Quartzite	1,6	tr	2,3	2,5	1,0	8,5	8,4	6,1	9,9	-	8,9	12,8	-	1,1	1,0	2,8
Sparite	-	tr	-	2,5	-	12,2	tr	14,8	6,6	5,2	-	-	-	9,3	35,6	tr
Micrite	-	1,0	27,1	7,5	-	24,8	-	2,5	10,3	13,1	5,1	1,0	-	4,1	14,9	33,0
Sericit-phyllite	-	-	-	-	-	-	-	-	-	-	-	-	25,0	-	-	-
Chlorite-phyllite	-	-	-	-	-	-	-	-	-	6,0	5,1	-	-	-	-	-
Mica-schist	-	-	-	-	-	-	-	-	tr	-	-	-	-	-	-	-
Calcitic fossil	-	-	-	1,0	-	-	-	tr	-	tr	-	-	-	3,0	tr	-
Opalic fossil	-	-	-	1,0	-	-	tr	-	-	tr	-	-	-	-	tr	-
Quartzite fossil	-	-	-	-	-	-	-	-	-	-	-	-	-	-	-	tr
Spicula	tr	-	-	-	tr	-	-	-	-	-	-	1,8	-	-	-	-

Table 5 Laecanius workshops amphorae

Measure of clasts /µm/	SPERATVS Emona No. 51	SPERATVS Aquileia 395	SPERATVS Aquileia 374	VIAT Aquileia 378	VIAT Savaria No. 150	VIAT Emona No. 52	VIAT Aquileia 389	VIAT Aquileia 443	CLAEK Aquileia 452	LAEKB Aquileia 449	LAEK Aquileia 400	LAEKB Poetovio No. 74	LAE Salla No. 96	LAEK Aquileia 375	LAI Aquileia 442	LAEK L. Aquileia 6177
Quartz	20-50	60-100	30-80	20-50	50-150	50-100	30-100	40-80	60-120	30-60	20-50	30-80	20-80	60-120	30-80	20-80
Othoclase	-	40-60	-	50-100	40-80	30-60	40-60	40-100	-	-	20-40	20-40	40-60	-	-	30-60
Plagioclase	-	10-20	20-60	40-60	30-50	-	40-80	40-80	60-150	20-50	-	10-40	40-50	60-150	40-60	20-80
Microcline	-	-	-	-	-	-	-	-	-	-	-	-	-	60-80	-	-
Muscovite	20-50	20-40	40-80	20-70	40-60	20-40	40-60	50-120	40-200	40-80	20-40	20-50	20-100	50-120	40-80	20-40
Biotite	30-50	20-50	40-80	20-50	20-30	40-60	30-60	40-80	40-80	30-60	20-40	30-60	20-80	50-150	40-80	20-50
Chlorite	-	-	-	-	-	-	-	-	-	-	-	-	-	40-120	-	-
Zoisite-clinozoisite	20-40	-	20-40	20-30	10-40	~30	20-30	30-50	20-40	20-40	-	~20	20-50	30-50	-	10-20
Tourmaline	-	-	30-50	20-40	20-40	-	-	20-30	-	-	-	~20	20-40	-	20-40	10-20
Rutile	-	-	-	-	-	~10	-	-	-	-	-	-	-	-	-	-
Apatite	20-60	-	-	-	-	-	-	-	-	-	-	-	-	-	-	-
Lazurite	-	-	-	-	-	-	-	-	-	-	-	~30	-	-	-	-
Zircon	30-40	-	10-30	10-20	5-10	~20	-	-	10-30	10-20	-	~10	10-20	15-30	-	-
Garnet	-	-	-	-	~10	-	20-30	-	10-20	-	-	-	10-20	-	-	-
Titanite	-	-	-	-	-	-	40-80	30-50	20-40	-	-	-	-	-	-	-
Opaque constituent	-	-	-	-	-	-	-	-	-	20-50	-	-	-	20-80	-	-
Epidote	-	-	-	-	~40	-	-	60-80	-	20-30	-	10-20	-	-	-	-
Quartzite	80-100	20-50	40-60	40-80	~100	50-90	40-250	40-80	60-200	-	30-60	40-80	-	60-100	40-80	20-50
Sparite	-	40-100	-	40-200	-	40-100	100-500	80-300	60-250	40-80	-	-	-	50-200	60-150	60-100
Micrite	-	40-80	50-150	50-150	-	60-120	-	40-120	60-150	60-90	40-60	50-80	-	50-200	80-250	100-240
Sericit-phyllit	-	-	-	-	-	-	-	-	-	-	-	-	20-150	-	-	-
Chlorite-phyllite	-	-	-	-	-	-	-	-	-	40-90	30-50	-	-	-	-	-
Mica-schist	-	-	-	-	-	-	-	-	100-150	-	-	-	-	-	-	-
Calcitic fossil	-	-	-	100-150	-	-	-	40-250	-	200-400	-	-	-	50-250	40-100	-
Opalic fossil	-	-	-	30-60	-	-	20-40	-	-	30-50	-	-	-	-	30-50	-
Quartzite fossil	-	-	-	-	-	-	-	-	-	-	-	-	-	-	-	20-50
Spicula	30-120	-	-	-	~150	-	-	-	-	-	-	~120	-	-	-	-

Table 5a Laecanius workshops amphorae

VIII. Amphora No. 96 LAE (Salla 83.5.91) cannot be assigned to any group. It contains abundant sericitephyllite (25%), which is not found in any other sherds. Carbonates are absent.

Remarks

The fragments of group I are similar to those assigned to group I of the Calvia Crispinilla workshop. But fragments from the Laecanius Bassus workshop have sparse quantities of carbonate.

Fragments of group II are similar to those of Dressel 2-4 group IV amphorae, and to the samples assigned to group II of the Calvia Crispinilla workshop.

The slight differences between groups IV and V are the rather higher quartz content and absence of calcite fossils in the former group. We cannot preclude the possibility that the clay was quarried from the same pit, in fossil-rich and fossil-poor beds. The samples of group VI are very similar to group II. The only differences are the following: the former group has no fossils and orthoclase was present in only one sample. Both groups of sherds could have been made from the same clay.

Workshop of Calvia Crispinilla

Amphorae from this workshop contain small quantities of mineral and rock fragments. Plagioclases are sparse (1.4 - 2.2%), if present at all. Orthoclase (3 - 14.5%) and carbonates (2 - 36.8%) are abundant in nearly all samples.

I. Abundant quartz (48.2 - 89.7%); orthoclase is present (less than 11.7%). Carbonates are almost absent.

CAL.CRISPINILLAE (Aquileia 400)

CAL.CRISPINILLAE (Aquileia 405)

No. 79 CAL.CRISPINILLAE (Poetovio 3579)

Sample Aquileia 405 contains a small amount of orthoclase. The Aquileia 400 sample shows sparse quartz and trace indications of carbonate. Hence these two samples are somewhat distinct from other members of the group, but the differences are not sufficient to justify a separate group.

CAL(VIA) CRISPINILLAE

	Aquileia 405	Aquileia 400	Aquileia 380	Emona No. 55	Emona No. 56	Poetovio No. 79	Poetovio No. 78	Poetovio No. 77	Mursella	Poreč 298
Matrix	95,2	95,3	91,0	96,8	93,2	96,8	95,1	92,8	92,8	95,1
Limonite	2,6	1,7	4,4	0,3	1,8	0,7	0,5	4,8	1,7	1,4
Clast	2,0	3,0	4,6	2,9	5,0	2,5	4,4	2,4	5,5	3,5
Distribution of clasts in %										
Quartz	89,7	48,2	39,4	52,3	52,3	60,4	28,0	38,6	47,3	64,2
Orthoclase	tr	5,3	-	-	3,0	9,9	7,0	14,5	6,9	7,9
Plagioclase	-	tr	2,2	-	-	tr	tr	-	tr	1,4
Muscovite	8,3	15,9	3,3	7,9	15,3	6,9	3,5	7,7	tr	12,5
Biotite	tr	17,3	8,0	3,0	8,8	18,8	7,0	11,1	4,2	10,4
Zoisite-clinozoisite	tr	tr	1,1	tr	1,0	2,0	-	-	1,1	tr
Tourmaline	tr	-	1,1	-	-	-	-	-	-	tr
Garnet	tr	-	-	-	-	-	-	-	-	-
Zircon	2,0	-	-	tr	tr	tr	1,0	-	tr	-
Apatite	-	-	-	-	-	-	1,0	tr	tr	-
Hypersthene	tr	-	-	-	-	-	-	-	-	-
Epidote	-	-	-	-	-	tr	-	-	-	-
Chlorite	-	-	tr	-	-	-	-	-	-	-
Quartzite	tr	12,0	-	-	5,0	2,0	2,9	19,6	3,8	3,6
Sparite	-	tr	36,6	36,8	12,8	-	8,4	8,5	32,8	-
Micrite	-	-	8,3	-	1,8	-	36,4	-	3,9	-
Sericit-phyllite	-	-	-	-	-	-	-	-	tr	-
Opalic fossil	-	1,3	-	-	-	-	-	-	-	-
Calcitic fossil	-	-	-	-	-	-	tr	tr	tr	-
Spicula	-	-	-	-	-	tr	4,8	tr	-	-

	Aquileia 405	Aquileia 400	Aquileia 380	Emona No. 55	Emona No. 56	Poetovio No. 79	Poetovio No. 78	Poetovio No. 77	Mursella	Poreč 298
Measure of clasts /μm/										
Quartz	40-60	50-60	40-100	20-50	30-80	2o-70	40-90	40-80	20-50	30-70
Orthoclase	60-80	30-40	-	-	40-60	20-50	~40	20-40	40-100	40-80
Plagioclase	-	40-50	40-80	-	-	~40	~50	-	40-60	40-80
Muscovite	20-30	30-40	40-60	30-60	40-120	30-80	20-40	30-40	40-100	40-120
Biotite	20-30	20-30	40-60	30-60	40-60	30-60	20-50	20-40	40-70	40-80
Zoisite-clinozoisite	20-40	10-20	20-40	10-20	15-25	10-20	-	-	30-70	10-20
Tourmaline	20-30	-	20-40	-	-	-	-	-	-	40-60
Garnet	10-20	-	-	-	-	-	-	-	-	-
Zirkon	10-20	-	-	10-20	15-25	~lo	~lo	-	~20	-
Apatite	-	-	-	-	-	-	~20	~20	~30	-
Hypersthene	30-50	-	-	-	-	-	-	-	-	-
Epidote	-	-	-	-	-	~15	-	-	-	-
Chlorite	-	-	40-100	-	-	-	-	-	-	-
Quartzite	20-30	30-50	-	-	40-80	20-40	20-40	40-80	40-60	60-120
Sparite	-	120-150	50-100	30-60	60-100	-	40-60	30-60	-	-
Micrite	-	-	50-100	-	60-80	-	80-150	-	40-70	-
Sericit-phyllite	-	-	-	-	-	-	-	-	~100	-
Opalic fossil	-	20-40	-	-	-	-	-	-	-	-
Calcitic fossil	-	-	-	-	-	-	~50	~40	50-100	-
Spicula	-	-	-	-	-	~40	~60	~20	-	-

Table 6 CAL.CRISPINILLAE stamp on amphorae

II. Quartz is common to abundant (28 - 52.3%), orthoclase is common (3 - 14-5%) and there are significant quantities of carbonates (1.8 - 36.4%).

No. 56 CAL.CRISPINILLAE (Emona F 7).

No. 77 CAL.CRISPINILLAE (Poetovio 3562).

No. 78 CAL.CRISPINILLAE (Poetovio 3548).

CAL.CRISPINILLAE (Mursella).

III. Quartz is common (39.4 - 52.3%) and carbonate abundant (8.3 - 36.8%), mostly sparite in both samples. Orthoclase and quartzite are absent.

No. 55 CAL.CRISPINILLAE (Emona F 562).

CAL.CRISPINILLAE (Aquileia 380).

Remarks

Group I is similar to group I of the Laecanius workshop, but the former does not contain carbonates.

Group II is also similar to group II of the Laecanius workshop, but the former contains more quartz.

Amphorae with CRISPINILI and CRI PIN stamps

Amphora No. 80 (Poetovio 3546) of the CRISPINILI stamped sherds is assigned to group II of the Crispinilla workshop, but its quartz content is somewhat lower, and there is more orthosclase. Other components are present in similar quantities. The Parentium 302 specimen can readily be assigned to group I of the Crispinilla workshop.

Other speciments with CRISPINILI stamps (Aquileia 407; Aquileia 5, and Poetovio 10932) and specimens with CRI PIN stamps (Emona 841 and Poetovio 10953) cannot be assigned to any of the groups.

	CRISPINILI						CRISPIN			CRISPINILI						CRISPIN		
	Aquileia 432	Aquileia 407	Aquileia 5	Poetovio No. 80	Poetovio No. 81	Poreč 302	Aquileia 447	Emona No. 57	Poetovio No. 82	Aquileia 432	Aquileia 407	Aquileia 5	Poetovio No. 80	Poetovio No. 81	Poreč 302	Aquileia 447	Emona No. 57	Poetovio No. 82
Matrix	96,9	93,0	94,6	96,9	94,6	94,4	94,2	96,9	93,5									
Limonite	0,6	2,2	-	1,0	4,3	2,6	1,2	0,4	2,7									
Clast	2,5	4,8	5,4	2,1	1,1	3,0	4,6	2,7	3,8									
Distribution of clasts in %										Measure of clsts /μm/								
Quartz	46,3	76,1	70,0	24,8	47,0	70,6	78,1	44,5	55,5	20-40	40-80	30-80	40-70	20-50	40-60	30-60	50-90	40-100
Orthoclase	4,9	-	-	24,8	tr	6,5	11,7	4,3	5,2	20-40	-	-	30-50	30-40	80-100	30-60	30-50	30-50
Plagioclase	9,7	3,1	4,1	tr	tr	-	-	1,4	6,3	30-50	40-80	30-60	~40	20-40	-	-	~30	40-60
Muscovite	17,1	4,6	5,0	20,0	6,0	10,8	1,9	13,2	11,1	10-30	30-60	30-60	30-60	30-40	40-100	30-60	30-60	20-50
Biotite	17,9	7,6	4,1	3,8	17,1	6,9	5,2	2,3	11,0	10-30	30-60	20-50	20-40	20-30	40-100	30-80	20-40	20-40
Zoisite-clinozoisite	-	1,0	1,0	-	1,0	tr	1,0	-	1,0	-	20-50	10-30	-	~10	30-40	20-40	-	~30
Epidote	-	-	-	1,9	-	-	-	-	-	-	-	-	10-25	-	-	-	-	-
Zircon	-	tr	-	-	tr	tr	-	1,4	tr	-	20-30	-	-	~10	~20	-	10-20	~10
Tourmaline	-	tr	tr	-	-	tr	1,1	-	2,1	-	20-40	10-30	-	-	~40	20-40	-	~20
Rutile	-	-	-	-	-	-	1,0	-	tr	-	-	-	-	-	-	20-40	-	~10
Apatite	-	tr	-	-	-	-	-	-	tr	-	20-60	-	-	-	-	-	-	~30
Opaque constituent	-	-	14,1	-	-	-	-	-	-	-	-	10-40	-	-	-	-	-	-
Quartzite	tr	-	-	11,4	2,0	5,2	-	2,2	7,8	20-40	-	-	30-50	30-40	40-50	-	20-60	40-60
Sparite	-	7,6	1,7	7,6	15,6	-	-	-	-	-	100-200	30-60	40-70	30-60	-	-	-	-
Micrite	-	-	-	3,8	9,2	-	-	-	tr	-	-	-	30-50	30-40	-	-	-	~60
Calcitic fossil	-	-	-	1,9	2,1	-	-	29,3	-	-	-	-	40-80	10-20	-	-	50-100	-
Opalic fossil	4,1	tr	-	-	-	-	-	-	-	40-60	40-80	-	-	-	-	-	-	-
Spicula	-	-	-	tr	-	-	-	1,4	-	-	-	-	~30	-	-	-	~40	-

Table 7 CRISPINILI and CRISPIN stamps on amphorae

Amphoras from Loron

Five groups were recognized and they are described below.

I. Abundant quartz (66.6 - 74.2%); carbonate is absent in specimens 1 and 2. Modest quantities of calcite fossils (6.5 - 7.8%). Orthoclase and plagioclase occur as well. Relatively higher incidence of accessory minerals (garnet, zircon, zoisite, tourmaline). Sparse quartzite (2.3 - 4%).

No. 57 CRI?PIN (Emona M 841) has calcite fossils.

II. Specimens 2 and 3. Quartz is common (37.1 - 40.1%) so too are carbonates (mostly sparite: 41.2 - 51.7%). Orthoclase and plagioclase are absent. An accessory mineral (zoisite) is present in one sample only.

This group is rather similar to group III of the Crispinilla workshop.

III. Quartz is abundant (54.8 - 68.4%). Orthoclase is present (traces to 3.2%) and plagioclase is sparse (1 - 2.6%). It has a high mica content (14.4 - 19.1% muscovite, and 14.8 - 15.9% biotite). Sparse quartz (traces to 4.4%). Carbonates are absent. Sherds 5 and 6 belong to this group.

This group can be assigned to group I of the Crispinilla workshop.

IV. Specimens 7 and 8 differ in the 21% sericite-quartzite content of the former. The other clasts are the same, so the two pieces have been assigned to the same group.

Common to abundant quartz (43.5 - 65.9%); medium amount of orthoclase (4.6 - 6.2%); plagioclases (1.5%). Relatively large sparse amounts of quartzite (9.2%). Carbonates are absent. This group is similar to the former so a connection can be postulated. It may be questioned whether specimens 7 and 8 belong to Group I of the Crispinilla workshop.

Quartzite content shows some differences, but both samples contain zoisite.

V. Specimens 9 and 10 from Loron are very similar. Quartz is common (46.4 - 48.9%); clase (2.1 - 5.8%) and plagioclase (1 - 3.2%) are also present. Relatively common too is quartzite (10%) and micrite (10%).

These sherds are similar to amphorae of group VII from the Laecanius workshop.

Amphorae with imperial stamps

These amphorae bear stamps of the emperors Domitian (IMP.AVG.GER.), Nerva (IMP.NERVAE.AVG) and of unspecified emperors (IMP). We also analysed samples for comparative purposes found not along the Amber Route but in Aquincum and Parentium (Porec).

Amphorae with IMP.AVG.GER. stamps

The Aquileia 381 amphora from the Amber Route is very similar to the samples of group V from the Laecanius workshop. Poetovio No. 83, Salla No. 102 and the Parentium 306 amphorae are similar to group I from the Crispinilla workshop, while Emona No. 70 and the Parentium No. 304 samples are similar to group II from the same workshop.

Amphorae with IMP. NERVAE.AVG stamps

The amphora Poetovio No. 84 cannot be assigned to any of our groups because of its extremely high phyllite content (61.2%).

The amphora Aquincum 41673 has been assigned to group V from the Laecanius workshop.

Amphora with IMP stamp

The amphora Aquincum 55.304 is related to group V from the Laecanius workshop.

Other Dressel 6B type amphorae

Two groups were recognized. They are described below.

I. Much quartz and sparse to medium amounts of plagioclase are characteristic of this group. K-feldspars are mostly represented by microcline, while sanidine and orthoclase were not found in all samples. Carbonates are abundant and calcite fossils can be seen in the thin sections (with one exception). Augite occurs in almost all amphorae in sparse quantities. Quartzite is abundant.

No. 66 L.IVN.PATIEN (Emona 64-214).

No. 67 L.IVN.PATIEN (Emona 1963).

No. 60 COSAE (Emona F 1932/OJ).

No. 62 APIC (Emona).

The amphora Emona No. 66 is less emphatically related to the other four amphorae (less quartzite and no K-feldspar), but the other components indicate this group.

IMP.AVG.GER

	Aquileia 381	Emona No. 70.	Poetovio No. 83.	Salla No. 102.	Aquincum 70.8.155.	Poreč 304	Poreč 306	Aquileia 381	Emona No. 70.	Poetovio No. 83.	Salla No. 102.	Aquincum 70.8.155.	Poreč 304	Poreč 306
Matrix	93,6	99,0	97,7	93,5	91,8	94,6	95,7							
Limonite	3,1	-	0,3	1,8	2,8	1,4	1,0							
Clast	3,3	1,0	2,0	4,7	5,4	4,0	3,3							
Distribution of clasts in %								Measure of clasts /µm/						
Quartz	40,4	24,3	94,0	77,6	41,1	45,4	48,3	30-50	20-40	40-70	40-80	10-60	30-60	10-60
Orthoclase	-	2,4	tr	7,1	-	10,9	10,1	-	20-30	~40	40-80	-	40-70	20-50
Plagioclase	1,3	4,7	-	tr	-	2,2	-	40-50	30-50	-	10-30	-	40-60	-
Muscovite	7,5	14,5	tr	7,2	2,7	6,7	16,9	10-30	20-30	20-30	60-100	40-100	25-40	30-50
Biotite	15,4	-	-	8,1	2,5	6,7	22,0	10-30	-	-	30-50	20-60	30-50	30-50
Zoisite-clinozoisite	-	-	-	tr	tr	tr	tr	-	-	-	20-30	20-30	10-20	10-20
Tourmaline	-	-	-	tr	-	tr	-	-	-	-	~40	-	~25	-
Rutile	tr	-	-	-	-	-	-	10-20	-	-	-	-	-	-
Apatite	-	-	-	-	-	tr	-	-	-	-	-	-	tr	-
Zircon	-	-	4,0	tr	tr	-	-	-	-	30-40	20-30	5-10	-	-
Garnet	-	-	-	tr	-	-	-	-	-	-	~40	-	-	-
Epidote	-	-	-	tr	tr	-	-	-	-	-	~30	~30	-	-
Quartzite	1,9	-	2,0	-	-	5,8	2,7	20-40	-	~30	-	-	60-80	30-40
Sparite	-	12,0	-	-	49,2	18,2	-	-	30-70	-	-	20-200	80-150	-
Mikrite	33,5	42,1	-	-	4,5	4,1	-	70-90	80-150	-	-	40-100	80-200	-
Sericit-phyllite	-	-	-	tr	-	-	-	-	-	-	60-100	-	-	-
Calcitic fossil	-	tr	-	-	-	-	-	-	~20	-	-	-	-	-

	IMP.NERVAE.AVG		IMP		IMP.NERVAE.AVG		IMP	
	Poetovio No. 84.	Aquincum 41673	Aquincum 1	Aquincum 55.3o4.	Poetovio No. 84.	Aquincum 41673	Aquincum 1	Aquincum 55.3o4.
Matrix	96,6	88,5	93,2	91,3				
Limonite	0,5	3,6	2,2	0,6				
Clast	2,9	7,9	4,6	8,1				
Distribution of clasts in %					Measure of clasts /µm/			
Quartz	17,2	60,9	24,0	40,9	30-60	20-70	40-120	40-100
Orthoclase	-	-	-	tr	-	-	-	60-80
Plagioclase	-	tr	tr	-	-	30-40	10-20	-
Muscovite	9,0	3,3	7,0	7,4	20-40	20-100	40-60	40-100
Biotite	9,6	tr	24,0	4,3	20-30	10-40	10-80	20-40
Zoisite-clinozisite	-	-	-	tr	-	-	-	30-50
Zircon	tr	tr	4,0	tr	~10	10-30	10-40	20-40
Garnet	-	-	-	tr	-	-	-	~30
Epidote	tr	-	-	-	~30	-	-	-
Sparite	2,0	19,3	20,0	29,6	30-50	10-100	40-40	30-100
Micrite	1,0	13,8	21,0	17,8	30-70	50-300	40-100	60-100
Phyllite	61,2	-	-	-	70-180	-	-	-
Sericit-phyllite	-	2,7	-	-	-	~100	-	-
Calcitic fossil	tr	-	-	tr	~60	-	-	50-300
Spicula	-	-	-	tr	-	-	-	20-200

Table 8 Imperial stamps on amphorae

II. This group can hardly be related to others. Abundant quartz, less plagioclase and common to abundant quantities of carbonates were found in the samples, with very different quantities of K-feldspar.

No. 49 CAMI (Nauportus).

No. 61 APICI (Emona 64-208).

No. 68 FLAV.FON (Emona 1350).

No. 85 PRIMI (Poetovio 3549).

The Emona No. 68 amphora contains conspicuous quantities of orthoclase; quartzite is sparse.

The Nauportus No. 49 sample shows significantly smaller grain size than other amphorae (50 - 60 m, and 50 - 500 m, in the other samples).

Dressel 6B amphorae without stamps

Specimens belonging to the Laecanius workshop:

IV. group: No. 114 (Salla 83.5.13).

V. group: No. 152 (Savaria 54.91.13).

VI. group: No. 151 (Savaria 64.15.2655).

VII. group: No. 127 (Salla 82.5.60).

Specimens belonging to the Crispinilla workshop are included in group II.

No. 113 (Salla 83.5.2).

No. 124 (Salla 83.5.57).

No. 126 (Salla 83.5.59).

No. 153 (Savaria 63.5.389).

Dressel 20

Globular amphorae from Baetica are indistinguishable as far as their clasts are concerned. But minor microscopic differences allow us to recognize two groups; they are described below.

I. Abundant quartz, less K-feldspar and plagioclase.

Augite is present. Quartz and carbonate are common to abundant. Very little microquartzite; sparse calcite fossils. Sparse to medium quantities of micaschist are characteristic of Dressel 20.

No. 217 Q.L.S (Poetovio 3531).

No. 218 G.M.M (Poetovio 3532).

II. Amphora sherd No. 220 (Salla 83.5.80) which has been assigned to this group is characterized by abundant quartz; K-feldspar is represented by microcline. There is no plagioclase, augite or microquartzite, to distinguish it from group I. The characteristic micaschist, quartzite, carbonate and calcitic fossils are present.

The following Dressel 20 sherds did not come from the Amber Route, but they were examined and shown to belong to group I.

L.Q.S	(Aquileia).
DOMS	(Aquincum 56.41.18).
FSCIM/NIANO	(Aquincum 55298).

Schörgendorfer 558 amphora

J. Ivâncsics

Examination of clastic fragments of Schörgendorfer 558 amphorae indicate they were produced in several different workshops. Most contain clastic material of volcanic origin (augite, Ti-augite, aegirine-augite, pigeonite, oxyamphibole and volcanic clasts). They must have been made in regions where these minerals and rocks are common. The sherds examined have been divided into carbonatic and carbonate-free groups. Of the carbonatic fragments

No. 279 (Salla 83.5.30) and

No. 277 (Poetovio 10951)

could have been made in Istrian workshops because they are strongly calcareous and contain many fossils. The following sherds are less certainly assigned to this group:

No. 280 (Salla 83.5.81)

No. 276 (Poetovio)

No. 284 (Salla 83.5.84)

No. 290 (Savaria 61.3.864) and

the amphora from Vindobona.

In most amphorae the carbonate content is not of primary clastic origin, but secondary, impregnation origin (pseudomorphs).

Those amphorae containing more or less clasts of magmatic origin were most probably made in Italy. They are listed below.

No. 281 (Salla 83.5.82).

No. 283 (Salla 83.5.83).

No. 286 (Savaria 54.321.26).

No. 287 (Savaria 54.321.27...).

No. 288 (Savaria 65.49.82).

No. 289 (Savaria).

	Savaria No. 287	Savaria No. 288	Savaria No. 289	Salla No. 283	Salla No. 281	Savaria No. 286	Salla No. 279	Salla No. 280	Poetovio No.276	Savaria No. 290	Salla No. 284	Vindobona	Poetovio No. 277	Celeia No. 275	Salla No. 285	Salla No. 282	Aquileia 3503	Aquincum
Matrix	62,6	70,5	81,4	70,9	79,0	74,1	75,1	72,7	83,3	78,0	79,7	82,4	36,3	72,4	75,0	71,0	49,5	36,0
Clast	37,4	29,5	18,6	29,1	21,0	25,9	24,9	27,3	16,7	22,0	20,3	17,6	63,7	27,6	25,0	29,0	50,5	64,0
Measure of clasts /μm/ average	27	36	30	16	27	45	48	26	38	17	21	23	26	21	30	35	42	15
Distribution of clasts in %																		
Quartz	76,4	76,6	30,8	72,0	70,0	46,9	73,6	53,5	65,0	54,0	36,8	46,8	36,1	59,3	52,0	35,0	37,5	64,8
Orthoclase	5,4	3,2	16,3	3,3	3,0	-	tr	8,0	3,4	3,0	16,2	9,6	0,6	tr	6,0	7,0	14,2	2,1
Plagioclase	5,8	5,1	33,3	5,1	8,0	4o,7	-	18,0	15,4	13,0	23,5	29,3	5,4	3,0	10,0	40,0	12,0	19,1
Sanidine	-	tr	-	-	tr	-	-	-	tr	-	-	-	-	-	-	-	-	-
Microcline	tr	-	0,7	-	1,0	tr	-	tr	tr	-	-	tr	-	-	-	-	0,3	-
Feldspar + quartz	0,9	0,9	-	0,5	-	-	-	-	-	-	-	-	-	-	-	-	-	-
Muscovite	1,7	3,1	2,7	5,0	4,0	0,6	1,3	5,0	2,5	10,0	2,1	1,4	8,3	7,1	6,0	1,0	3,7	2,0
Biotite	4,2	7,3	5,5	8,6	6,0	7,8	6,7	3,5	8,3	13,0	4,9	4,4	tr	2,1	3,0	6,0	3,6	9,3
Clinozoisite	0,9	0,5	1,2	0,5	1,0	-	-	tr	-	tr	1,4	1,5	-	-	-	tr	tr	tr
Tourmaline	-	-	-	tr	-	-	-	-	-	1,0	-	-	-	-	-	-	-	-
Chlorite	tr	0,2	tr	0,3	tr	-	-	-	-	-	tr	tr	-	-	-	-	-	-
Apatite	-	-	tr	-	-	-	-	-	-	-	tr	tr	-	-	-	-	-	-
Analcite	-	-	-	-	-	tr	-	-	-	-	-	-	-	-	-	-	-	-
Zircon	tr	tr	tr	tr	tr	tr	-	tr	-	-	tr	tr	-	-	-	-	tr	tr
Titanite	-	-	tr	-	-	-	-	tr	-	-	-	-	-	-	-	-	-	-
Rutile	-	tr	-	0,3	tr	-	-	tr	-	-	tr	0,3	-	-	-	-	-	-
Garnet	0,3	tr	tr	0,5	tr	-	tr	tr	-	1,0	tr	tr	-	-	-	-	-	-
Augite	0,5	1,3	1,2	0,3	1,0	tr	tr	tr	-	-	-	-	-	-	tr	-	-	0,2
Ti-augite	-	-	-	tr	-	-	-	-	-	-	-	-	-	-	-	-	-	-
Aegirine-augite	-	-	-	-	-	-	-	-	tr	-	-	-	-	-	-	-	-	-
Epidote	-	-	-	tr	tr	-	-	-	-	-	tr	-	-	-	-	-	-	-
Dolomite	-	-	-	-	tr	tr	-	0,5	-	tr	-	0,3	-	-	-	-	-	-
Phyllite	-	-	-	tr	tr	-	-	tr	tr	tr	-	-	-	-	1,0	-	-	-
Granophire	-	-	-	-	-	-	-	tr	-	-	tr	-	-	-	-	-	-	-
Pigeonit	tr	-	tr	tr	-	-	-	-	-	-	-	-	-	-	-	-	-	-
Corundum	-	-	-	-	-	-	-	-	-	-	tr	-	-	-	-	-	tr	-
Opaque constituent	0,3	-	0,5	-	-	tr	0,6	0,5	1,3	-	-	-	-	-	2,0	1,0	tr	tr
Vulcanic rock	-	-	-	-	tr	tr	-	-	-	-	-	tr	-	-	-	-	-	-
Limonite granulate	2,3	1,3	5,4	2,7	3,0	0,9	1,9	8,0	0,2	2,0	7,9	2,9	7,9	5,2	5,0	3,0	1,0	tr
Quartzite	1,3	0,5	2,4	0,9	2,0	1,9	2,6	2,0	-	2,0	2,1	2,0	-	23,3	8,0	7,0	tr	2,5
Carbonate	-	-	-	-	-	1,2	13,0	tr	0,1	tr	5,1	1,5	29,6	-	-	-	20,4	-
Mudstone	-	tr	tr	tr	1,0	tr	-	1,0	3,8	-	tr	-	tr	-	-	-	tr	tr
Claystone	-	-	-	-	-	-	-	-	-	1,0	tr	-	-	-	7,0	tr	tr	-
Oxyhornblende	-	tr	tr	-	-	-	-	-	-	-	-	-	-	-	-	-	-	-
Microfauna	-	-	-	tr	tr	-	0,3	-	tr	tr	tr	tr	12,1	-	-	-	7,3	-

Table 9 Schörgendorfer 558 amphorae

Sherds not mentioned above may form a distinct, third group.

Examination of the clastic components in Schörgendorfer 558 type amphorae - though we have to bear in mind the limitations of such investigations - indicate that a limited number were made in the Laecanius and Crispinilla workshops. Most come from workshops somewhere in central Italy.

X-ray powder diffraction analyses

T. Weisburg & G. Papp

1. Introduction

The fabric of amphorae (as coarse ceramics) contains larger grains, readily detectable with a hand lens (minerals, rock and fossil fragments) and a groundmass, not amenable to study by the same. Both constituent types have their role in the description of the materials that were used to make amphorae, and in any determination of the firing conditions.

X-ray powder diffraction methods are best suited for characterization of the groundmass.

2. Results of the analyses (arranged by amphora types and workshops)

Dressel 2-4

a. Quartz is dominant, with a significant amount of feldspars (plagioclase > K-feldspar). No calcite and mica.

No. 17 (Salla 83.5.105).

No. 19 (Salla 83.5.135).

No relationships with other groups.

b. Quartz is dominant, with significant quantities of feldspar and "mica". No calcite. Two sub-groups were identified, based on the ration of the two feldspars.

plagioclase > K-feldspar

No. 26 (Savaria 74.4.1370)

No. 7 (Poetovio)

Related to Aquileia 447 CRIƧPIN.

K-feldspar > plagioclase

No. 31 (Savaria 77.2.1811).

c. Quartz and calcite are dominant, with significant quantities of feldspars (plagioclase > K-feldspar) and "mica".

No. 12 (Poetovio R.500).

Related to five unstamped Dressel 6B/b group and to a sherd stamped CAL.CRISPINILLAE from Mursella.

d. Quartz and pyroxene are dominant, with unidentifiable feldspar.

No. 20 (Salla 83.5.106).

No. 10 (Poetovio 3685).

Related to C.LAECANIVS B: Aquileia 395 SPERATVS; Aquileia 389 VIAT and Aquileia 405..CRISPINILLAE.

Dressel 5

Besides the dominant quartz there are significant quantities of calcite and plagioclase. There is no K-feldspar and "mica".

No. 45 (Savaria 69.6.6).

No. 44 (Poetovio).

Readily distinguished from all other samples.

Dressel 6B

Workshop of Laecanius Bassus

a. Quartz is dominant, with significant quantities of plagioclase and "mica".

Aquileia 433 C.LAEK; BAR

Related to CRISPINILI 5 Aquileia; Aquileia 381, IMP.AVG. GER.

b. Quartz is dominant, with significant quantities of plagioclase and "mica". Less calcite.

Aquileia 400 LAEK.

Aquileia 376 C LAEK BASSI.

Aquincum 56.63.21, lower part of body C.LAEK.B; BAR.

Aquincum 56.63.21, body-neck C.LAEK.B; BAR.

Related to Aquileia 432 CRISPINILI

c. As (a), but with significant quantities of calcite.

Aquileia 399 ADEL.

No. 97. (Salla. 84.1.2. FELIX.PE.)

Aquileia 375 LAE.

Aquileia 374 SPER.

No. 96. (Salla 83.5.91. LAE.)

Related to two sherds with imperial stamps:

Aquincum 41673 (IMP.NERVAE.AVG) and 70.8.155. (IMP.AVG.GER)

d. As (a), but with calcite dominant.

Aquileia 391 CRESCENTIS

Aquileia 377 PIERI.

Aquileia 379 HER.

Aquileia 442 LAI.

e. As (b), but with characteristic K-feldspar (plagioclase > K-feldspar).

Aquileia 373 HOM.

Aquileia 452 CLAEK.

Aquileia 6177 LAE L.

Aquileia 378 VIAT.

No. 107. (Salla FEL.TVR 83.5.90) with sparse haematite.

f. Dominant quartz, with significant quantities of pyroxene and unspecifiable feldspar.

Aquileia 395 SPERATVS

Aquileia 389 VIAT.

Related to Dressel 2-4/d group, Aquileia 405 (CAL.CRISPINILLAE.)

g. Dominant quartz, with significant quantities of plagioclase, "mica" and gypsum, less calcite and K-feldspar.

Aquileia 430 LAEK HER.

Amphorae with the CAL. CRISPINILLAE stamp

a. Quartz is dominant with significant quantities of pyroxene and unspecifiable feldspar.

Aquileia 405

Related to two Dressel 2-4/d group, and to Laeconius Bassus/f group.

b. Quartz is dominant with significant quantities of feldspar (plagioclase > K-feldspar), less "mica" and calcite.

Aquileia 400

c. As (b), but with significant quantities of calcite and "mica".

Aquileia 380

Related to Laeconius Bassus/e.

Amphorae with CRISPINILI stamp

a. Dominant quartz, less plagioclase.

Aquileia 407

Related to Dressel 2-4 No. 16 (Salla 83.5.104.)

b. As (a), but with significant quantities of "mica".

Aquileia 5 CRISPINILI

Related to Laeconius Bassus/a; imperial stamp: Aquileia 381 IMP. AVG.GER.

Amphora with CRI PIN stamp

a. Quartz is dominant with significant quantities of feldspar (plagioclase > K-feldspar) and "mica".

Aquileia 447

Related to two Dressel 2-4/b group.

Amphorae with imperial stamps

IMP. AVG. GER.

a. Dominant quartz, with significant quantities of calcite, plagioclase, and "mica".

   Aquincum 70.8.155.

   Related to Dressel 6B with imperial stamp: Aquincum 41673 (IMP. NERVAE.AVG); Laeconius Bassus/c group.

b. Dominant quartz, with signficant quantities of gypsum (?)

   No. 102 (Salla 83.5.7 (...AVGG...))

   Unrelated to other groups.

c. Quartz is dominant, with significant quantities of plagioclase and "mica" Aquileia 381.

   Related Laeconius Bassus/a group, Aquileia 5 (CRISPINILI)

IMP. NERVAE. AVG.

The same as IMP. AVG. GER. /a

   Aquincum 41673

IMP. DOMITII

Dominant quartz, with significant quantities of plagioclase and "mica", less K-feldspar and calcite.

   Savaria 60.1.139.

   Related to two unstamped sherds: No. 152. (Savaria 54.91.13) and No. 102. (Salla 83.5.7.)

Unstamped amphorae

a. Dominant quartz, with significant quantities of plagioclase and "mica", less K-feldspar and calcite.

   No. 152. (Savaria 54.91.13.)

   No. 98. (Salla 83.5.77.)

   Related to Dressel 6B stamped No. 154. (Savaria IMP. DOMITII.)

b. Dominant quartz, with significant quantities of plagioclase and "mica", less K-feldspar.

No. 153. (Savaria 63.5.389.)

No. 113. (Salla 83.5.2.)

No. 114. (Salla 83.5.13.)

No. 124. (Salla 83.5.57.)

No. 126. (Salla 83.5.59.)

Related to Dressel 2-4/c group, and to an amphora from Mursella (CAL.CRISPINILLAE).

Conclusions

The X-ray powder diffraction analyses revealed the mineralogical composition of the groundmass of the amphorae; this enabled us to recognize several groups among the material. But, for a sound interpretation of these results, one must turn to the polarizing microscope.

Our preliminary investigations revealed that there are differences among the amphora types. Dressel 5 could be distinguished from all other specimens; Dressel 2-4 types are also characterized by particular minaralogical compositions. The latter sometimes overlap with some of the specimens from the Dressel 6B workshops.

Because most of the examined samples are from the Dressel 6B workshop of C. Laecanius Bassus, this remains the best known group.

We can only provide initial data on the conditions of firing. It seems certain that it was undertaken at relatively low temperatures, as indicated by the presence of (sometimes relatively significant) quantities of primary $CaCO_3$ (it dissociates at 950^oC). But, some of the samples contain pyroxene (considered as an alteration product), which would indicate somewhat higher temperatures. The differences in temperatures may be due to different firing technologies, or to different temperatures within the kiln (centre and margins).

Other methods (e.g. transmission electron microscope) might give more precise data on the firing temperatures, by addressing themselves to early alteration products, which are difficult or impossible of resolution by X-ray powder diffractometry.

J. Iváncsics

X-ray diffraction (Schörgendorfer 558 amphora)

In addition to the large amount of vitreous (not crystallized) material, X-ray diffraction studies have shown well-crystallized, main clastic components and microlite crystals in the matrix, undetectable by microscope.

The Ca and Mg content of the clay reacted with quartz during firing and formed diopside, a Ca-Mg silicate. This is a frequent component of fired ceramics and in scoria with not too high Fe content.

Mica components (muscovite, biotite) frequently observed by the microscope suffer disintegration in their crystal lattice, resulting in X-ray amorphous behaviour without any reflections.

Quartz and feldspar are the main representatives of clastic components. X-ray studies enabled us to recognize very small components, which are too small for microscopic determination.

No. 286 (Savaria 54.321.26)

Ascertained components: quartz (common), K-feldspar (little), plagioclase (little), calcite (sparse), diopside (sparse) and mica (sparse).

No. 287 (Savaria 54.321.28)

Ascertained components: quartz (common), plagioclase (little), K-feldspar (little), diopside (sparse) and mica (sparse).

No. 288 (Savaria 65.49.82)

Ascertained components: quartz (common), plagioclase (little), diopside (little), K-feldspar (sparse) and mica (sparse).

No. 289 (Savaria)

Ascertained components: quartz (little), diopside (little), plagioclase (little), K-feldspar (sparse) and mica (sparse).

No. 279 (Salla 83.5.30)

Ascertained components: quartz (abundant), plagioclase (little), diopside (sparse), K-feldspar (sparse) and mica (sparse).

No. 280 (Salla 83.5.81)

Ascertained components: quartz (common), plagioclase (little), K-feldspar (little), diopside (little), mica (sparse) and calcite (sparse).

No. 284 (Salla 83.5.84)

Ascertained components: quartz (common), plagioclase (little), diopside (little), K-feldspar (little) and mica (sparse).

No. 283 (Salla 83.5.83)

Ascertained components: quartz (common), plagioclase (little), K-feldspar (little), diopside (sparse), calcite (sparse) and mica (sparse).

VINDOBONA

Ascertained components: quartz (common), diopside (little) and K-feldspar (little).

Aquileia 3503.290.

Ascertained components: quartz (abundant), calcite (common), plagioclase (little), K-feldspar (little), diopside (sparse) and mica (sparse).

Aquincum COC PHE E

Ascertained components: quartz (common), plagioclase (little), diopside (little), K-feldspar (sparse) and mica (sparse).

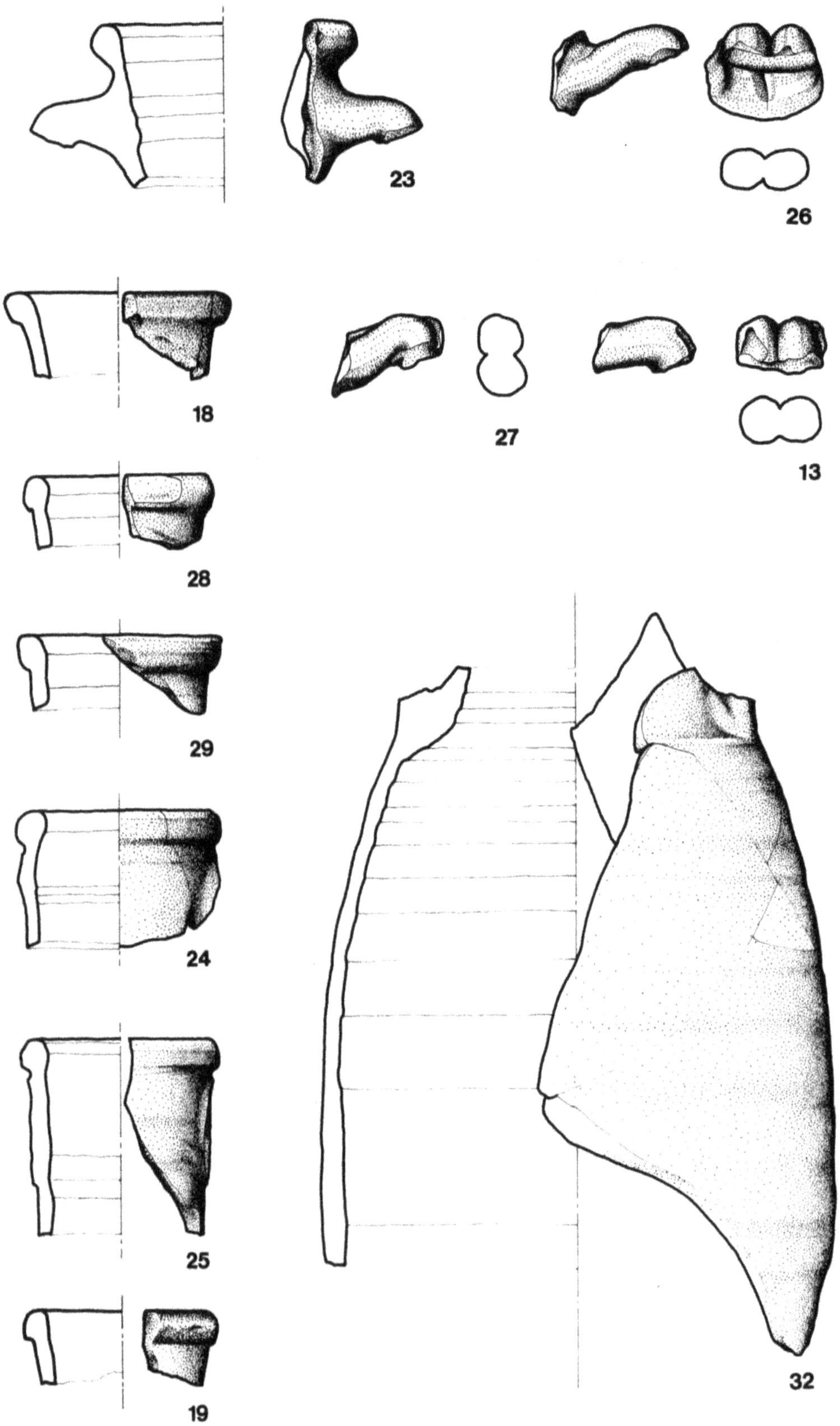

Figure 15 Dressel 2-4 amphorae (nos. 13, 18, 19) from Salla and (nos. 23-29, 32) from Savaria

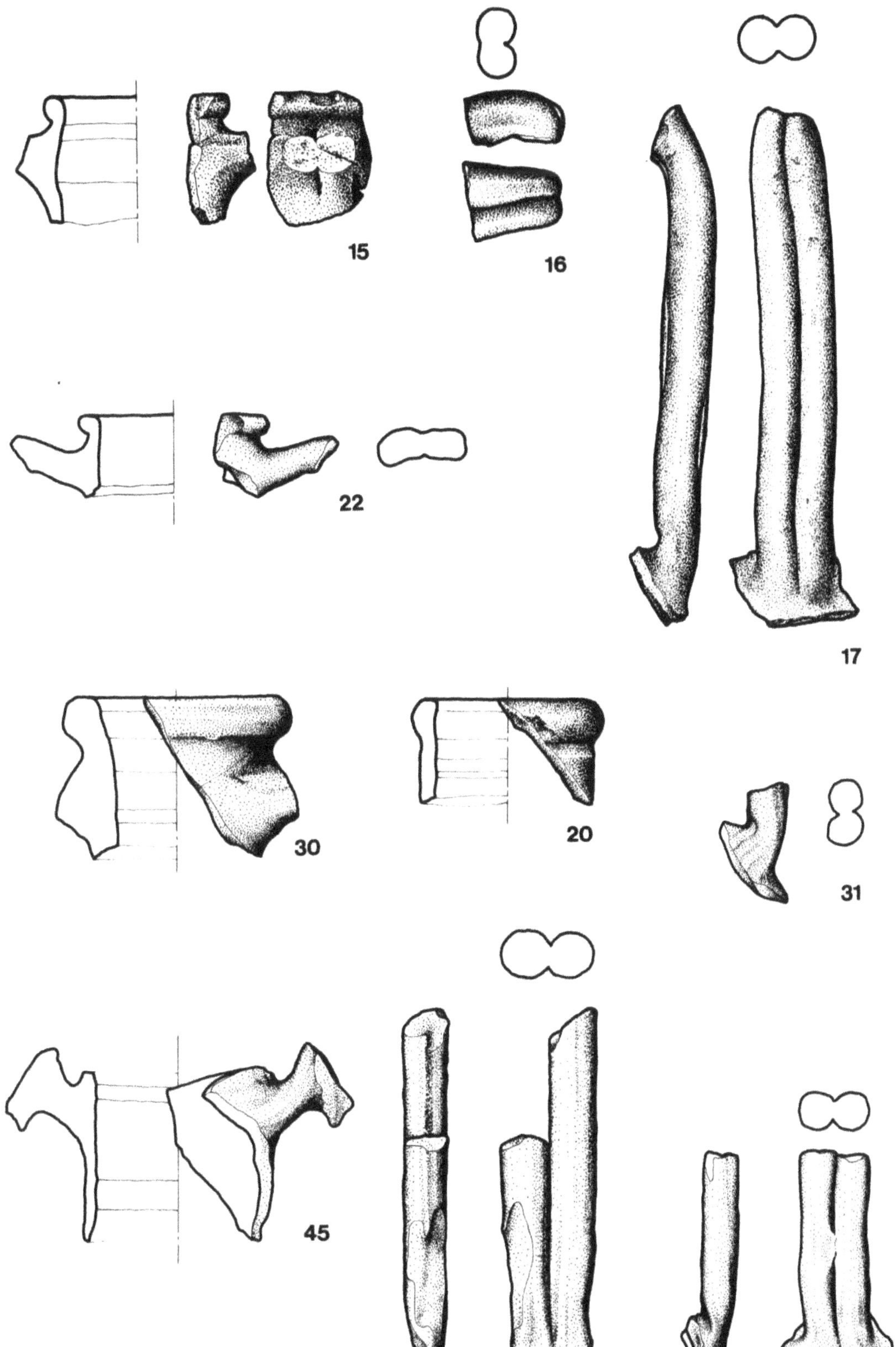

Figure 16 Dressel 2-4 amphorae (nos. 14-17, 20-21) from Salla, (no. 22) from Ad Arrabonem, (30-31) from Savaria, and Dressel 5 amphora (no. 45) from Savaria

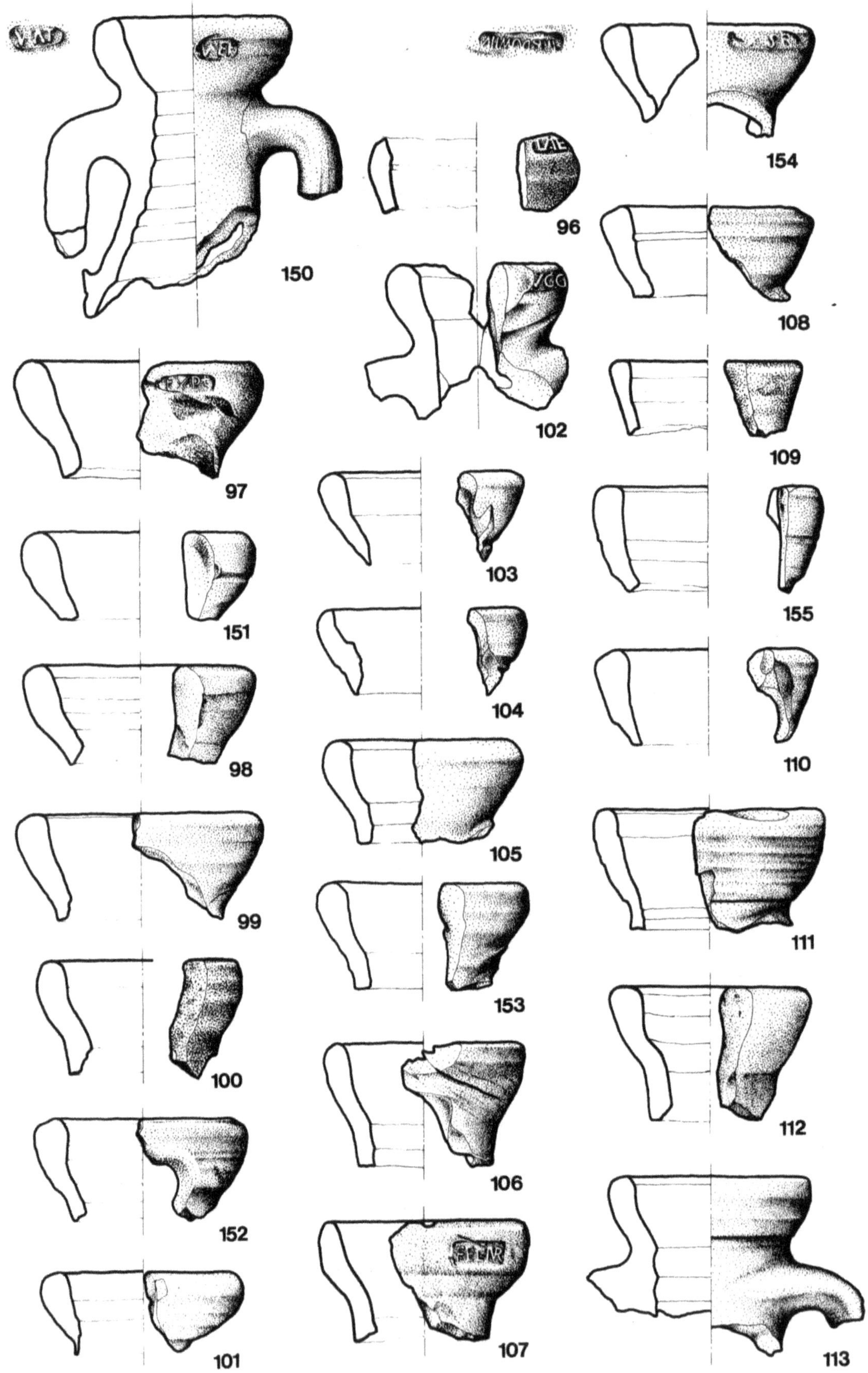

Figure 17 Dressel 6B amphorae (nos. 96–113) Salla and (nos. 150–155) Savaria

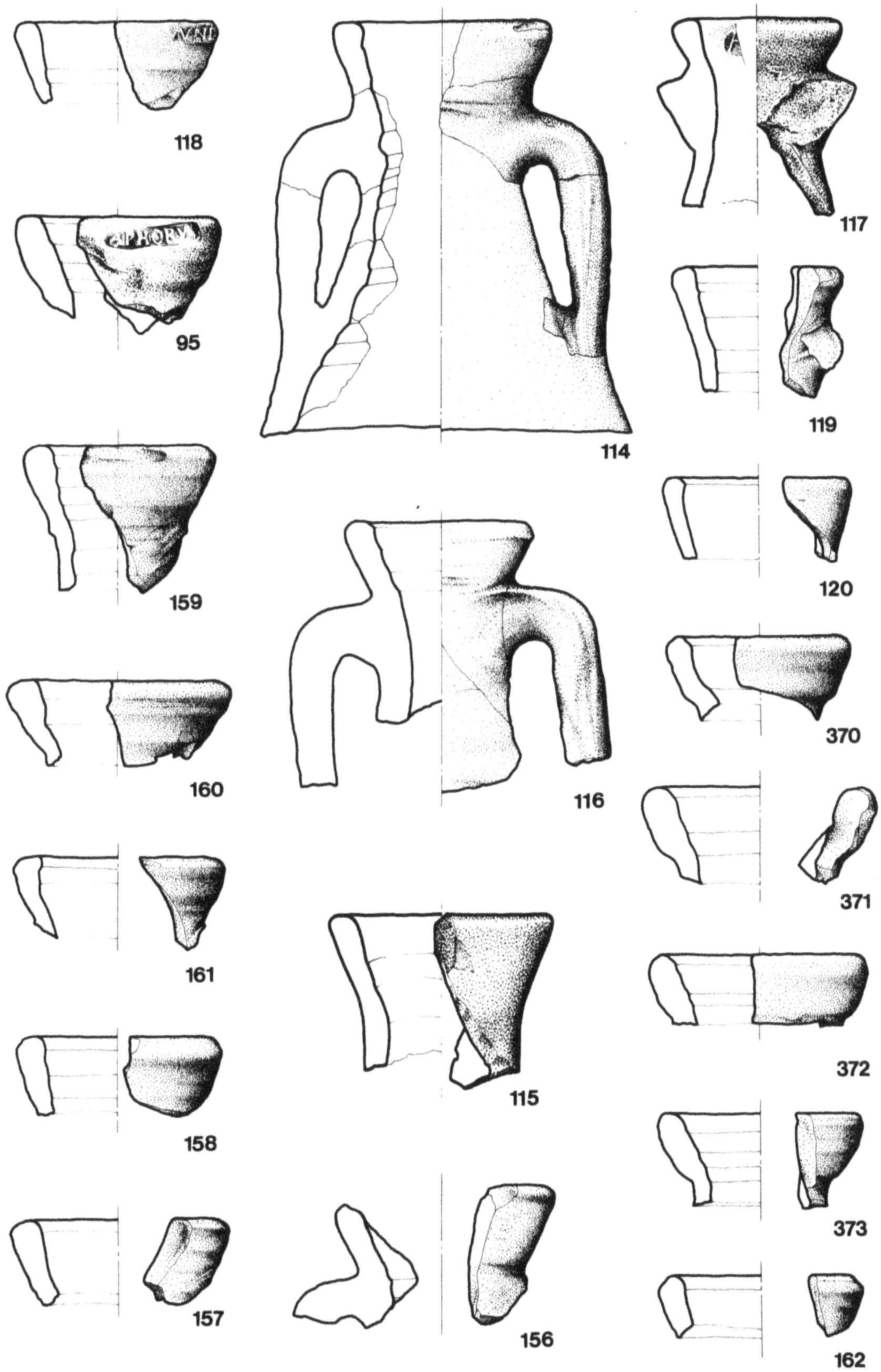

Figure 18 Dressel 6B amphorae (nos. 95, 114-120) Salla and (nos. 156-162, 370-373) Savaria

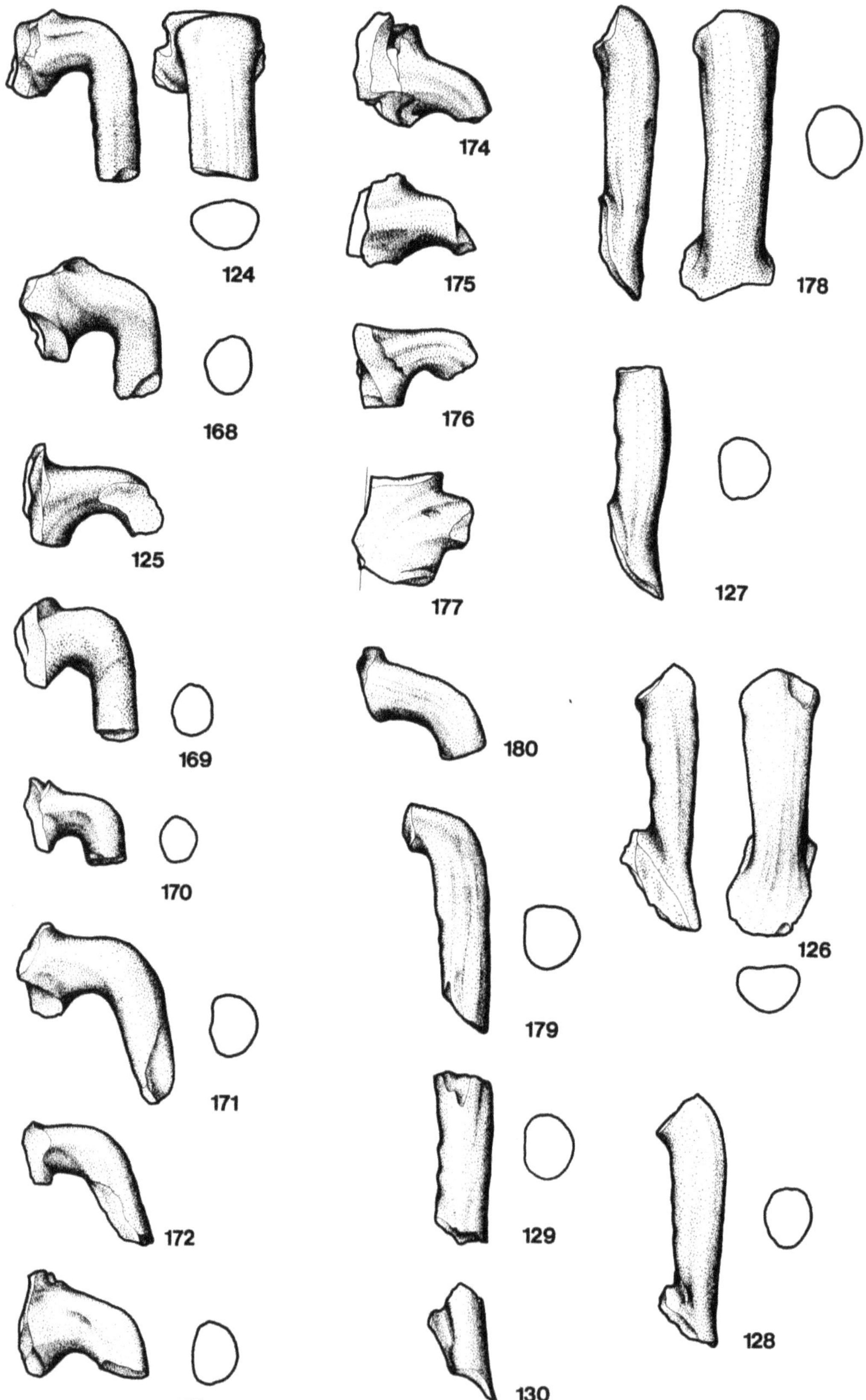

Figure 19 Dressel 6B amphora handles (nos. 124-130) Salla and (nos. 168-180) Savaria

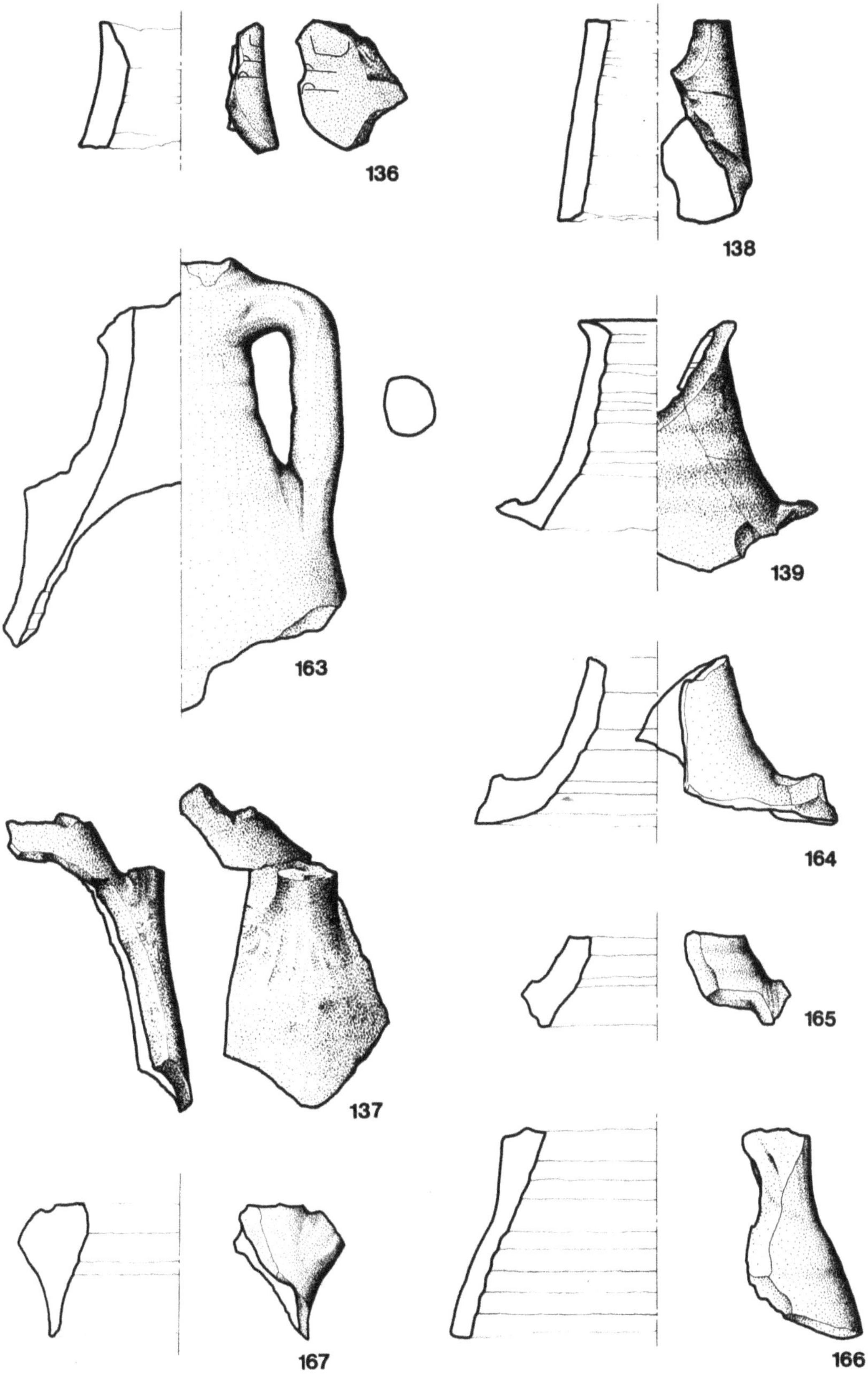

Figure 20 Dressel 6B amphorae (nos. 136-139) Salla and (no. 163-167) Savaria

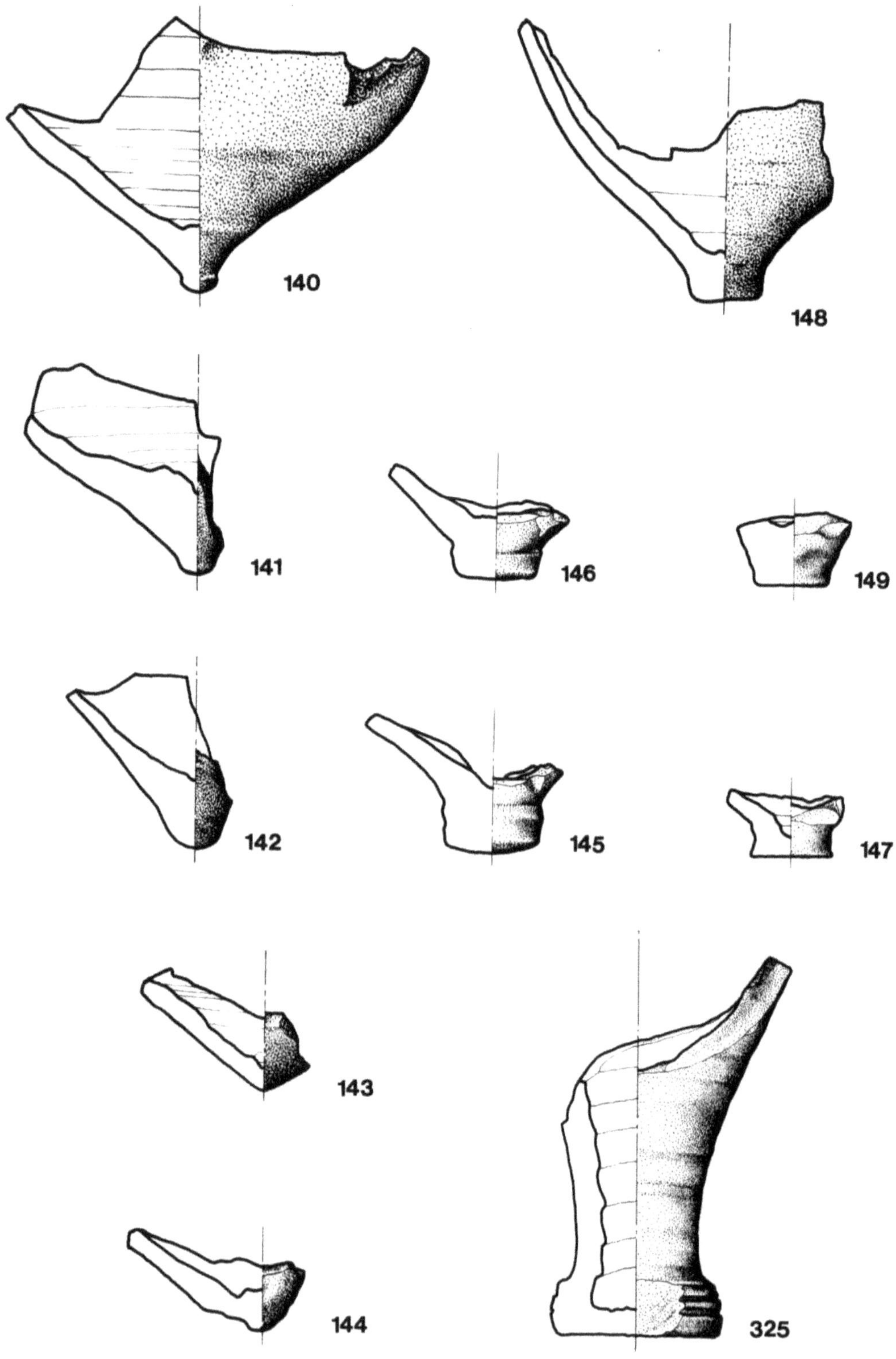

Figure 21 Dressel 6B amphora bases (nos. 140-149) Salla and an unidentified amphora base (no. 325) Savaria

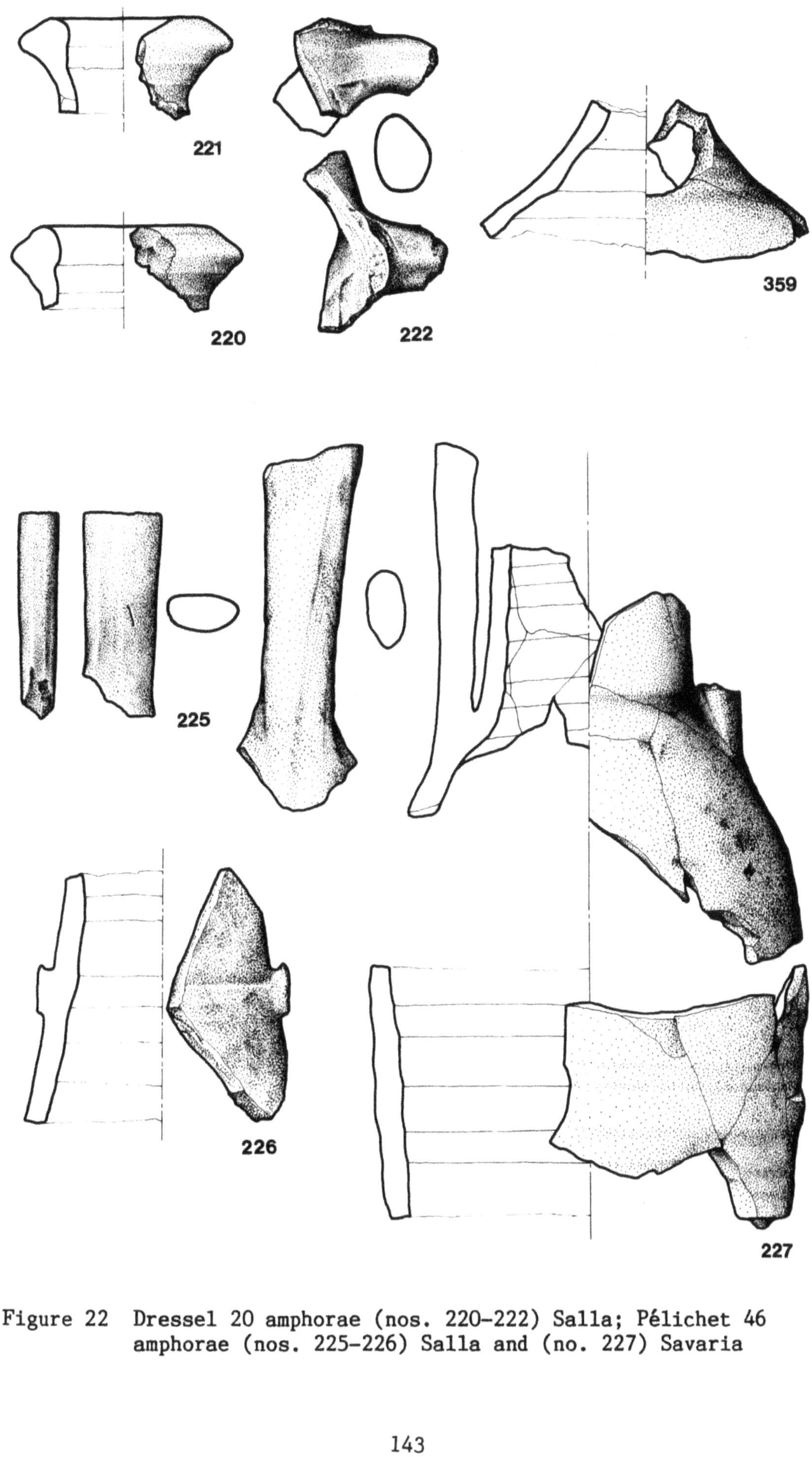

Figure 22 Dressel 20 amphorae (nos. 220-222) Salla; Pélichet 46 amphorae (nos. 225-226) Salla and (no. 227) Savaria

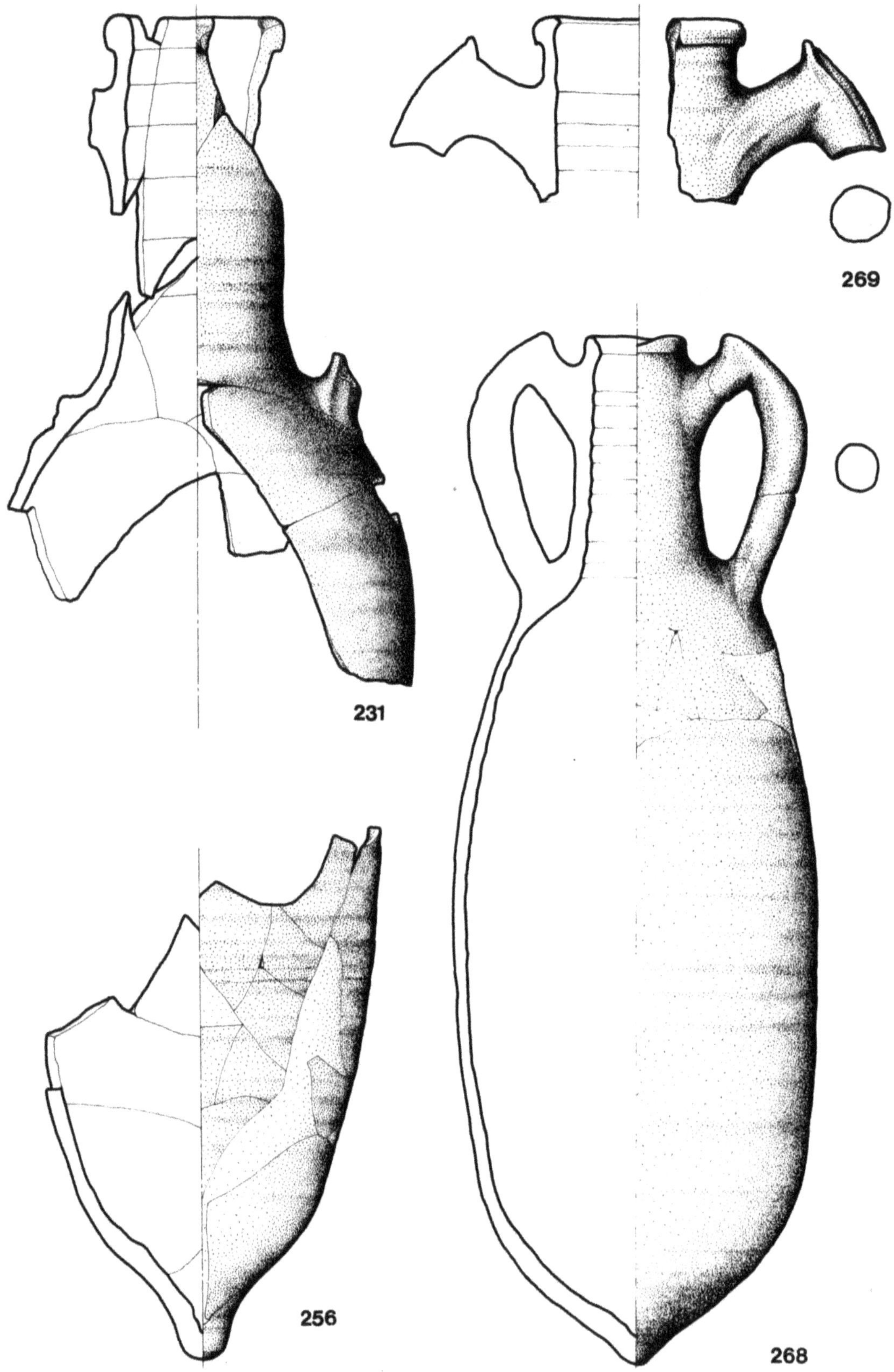

Figure 23 Rhodian amphorae (no. 231) Salla, (no. 256) Savaria and (nos. 268-269) Scarbantia

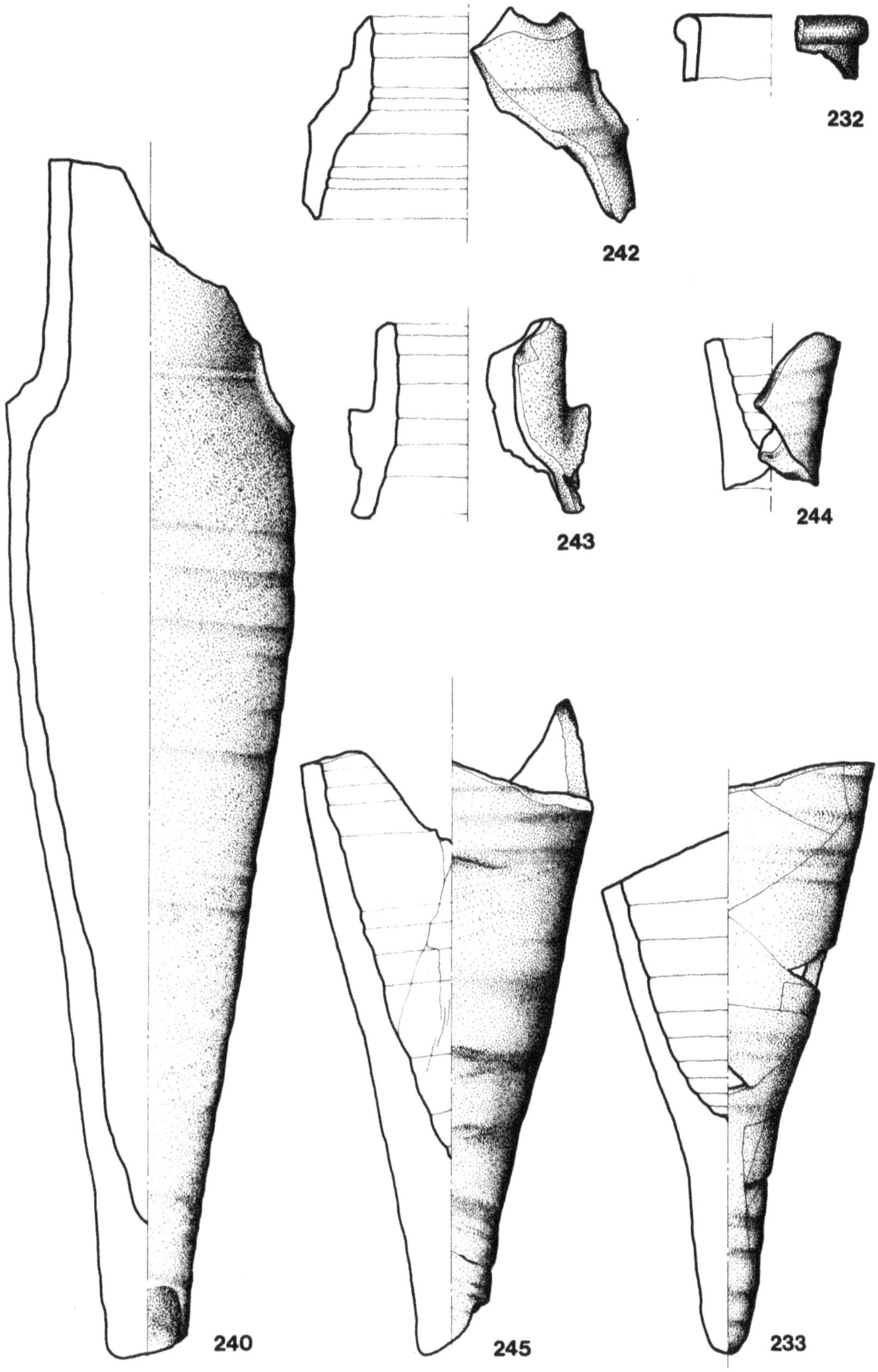

Figure 24 Rhodian amphorae (nos. 232-233) Salla and (nos. 240, 242-245) Savaria

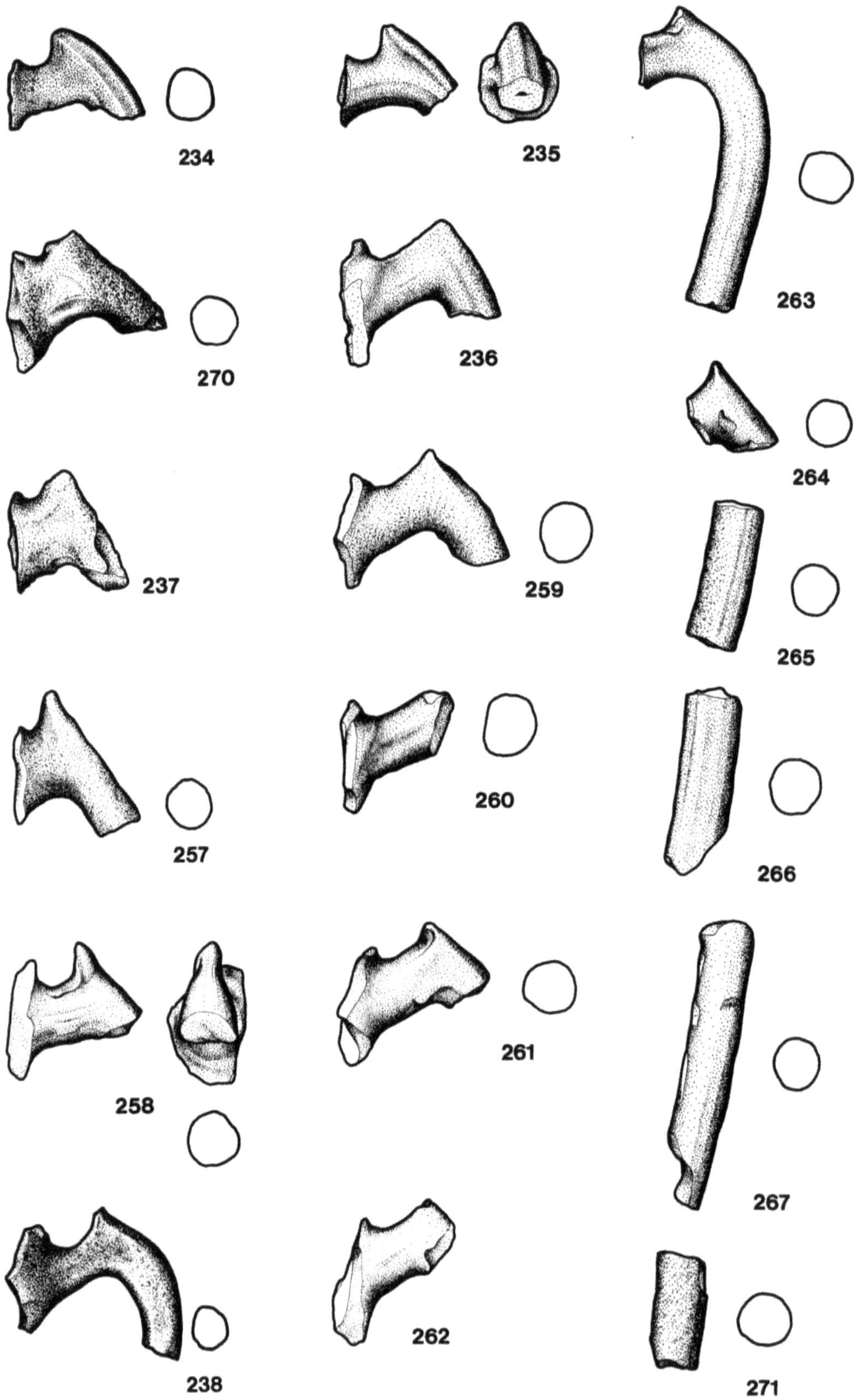

Figure 25 Rhodian amphora handles (nos. 234-238) Salla, (no. 257-267) Savaria and (nos. 267-271) Scarbantia

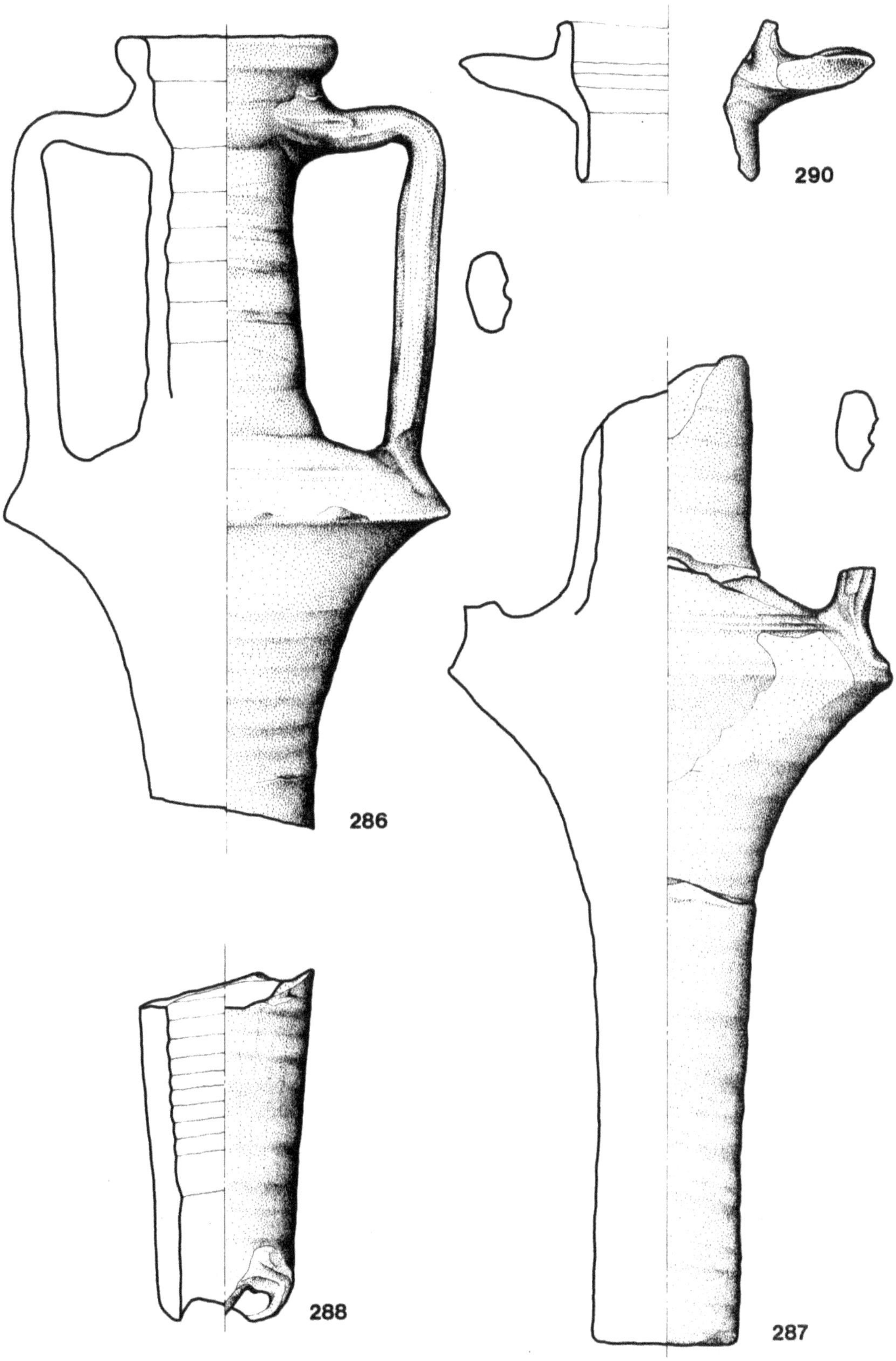

Figure 26 Schörgendorfer 558 amphorae (nos. 286-288, 290) Savaria

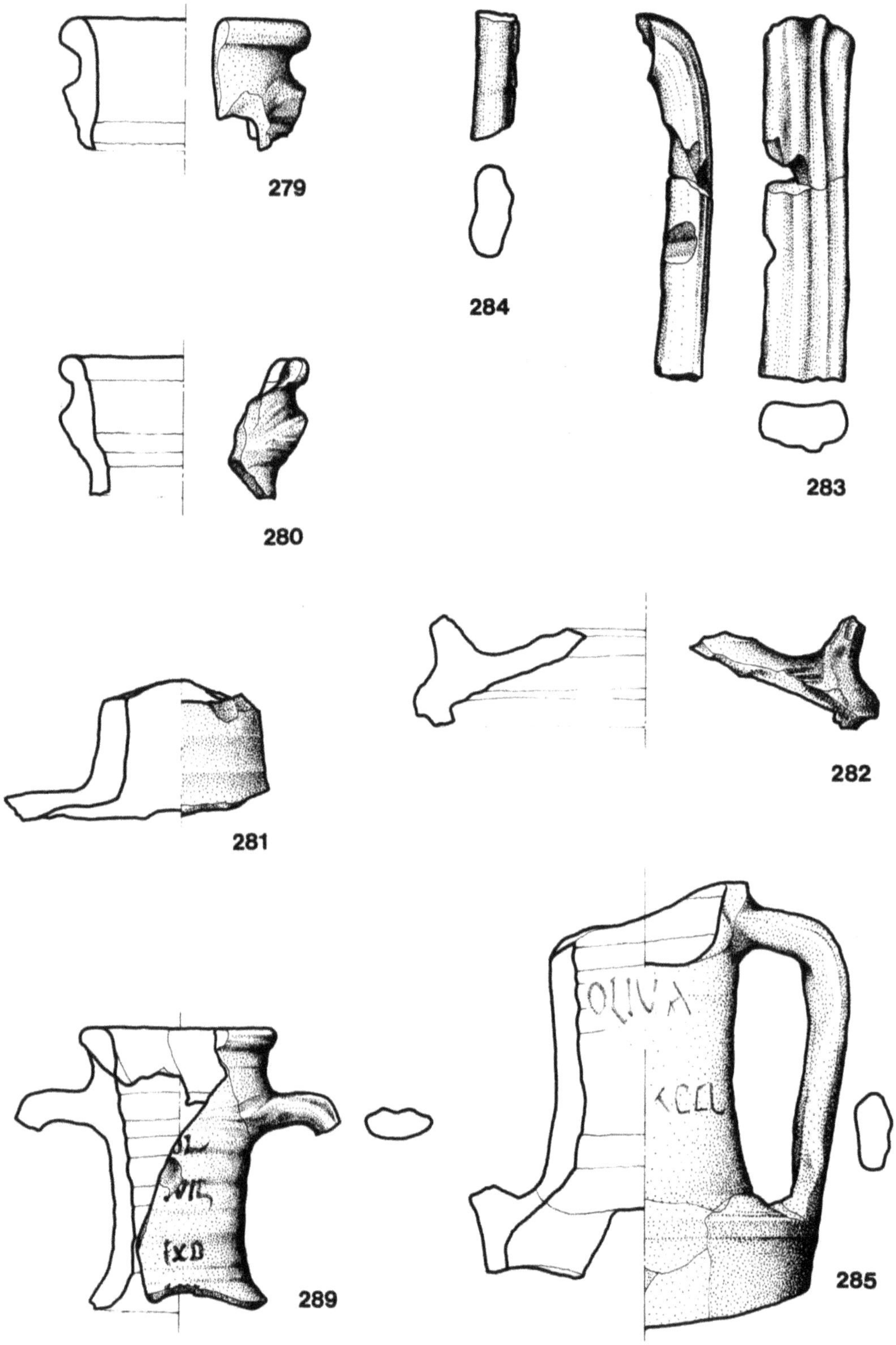

Figure 27 Schörgendorfer 558 amphorae (nos. 279-285) Salla and (no. 289) Savaria

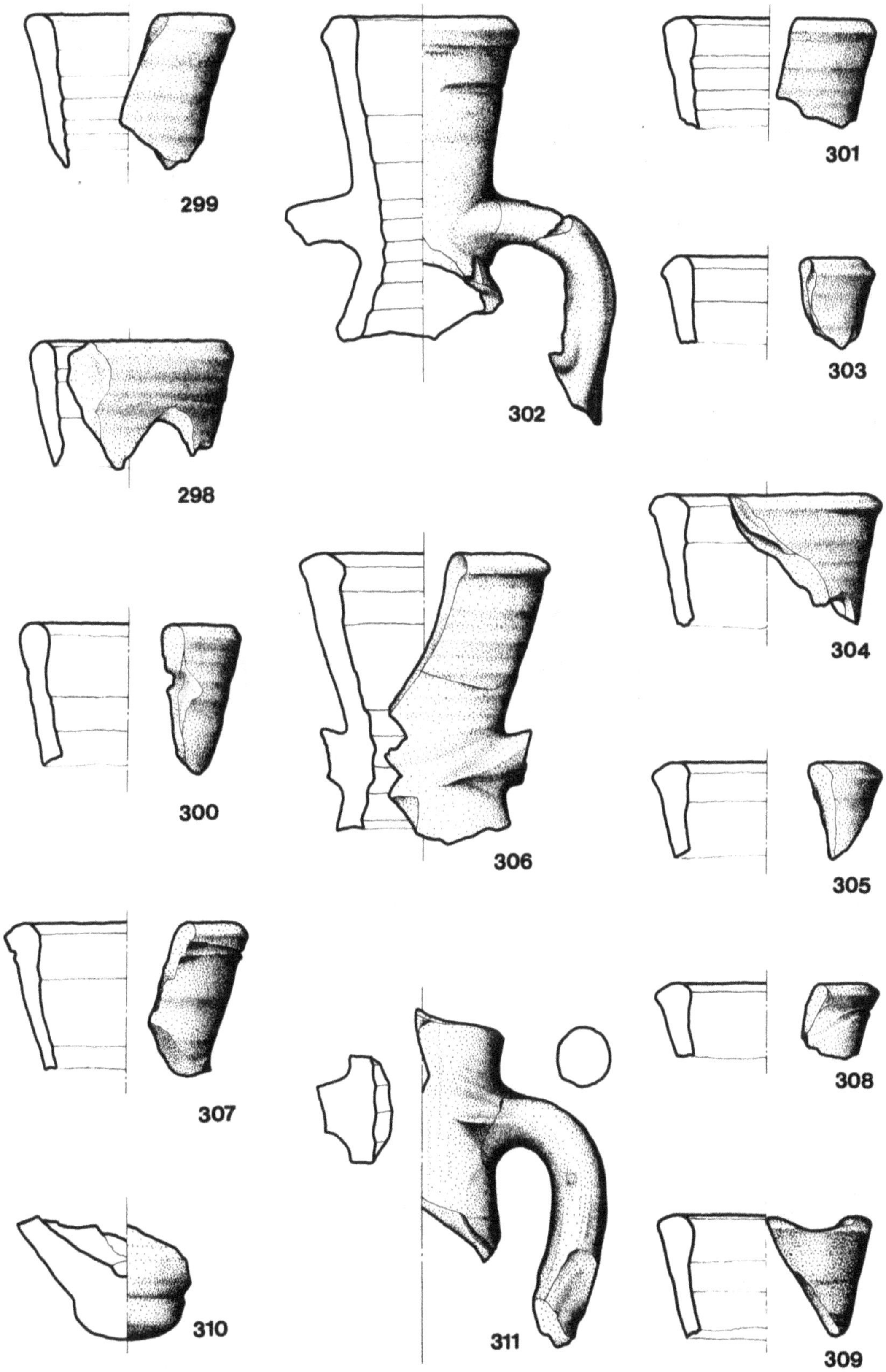

Figure 28 Portorecanatian amphorae (nos. 298-310) Salla and (no. 311) Savaria

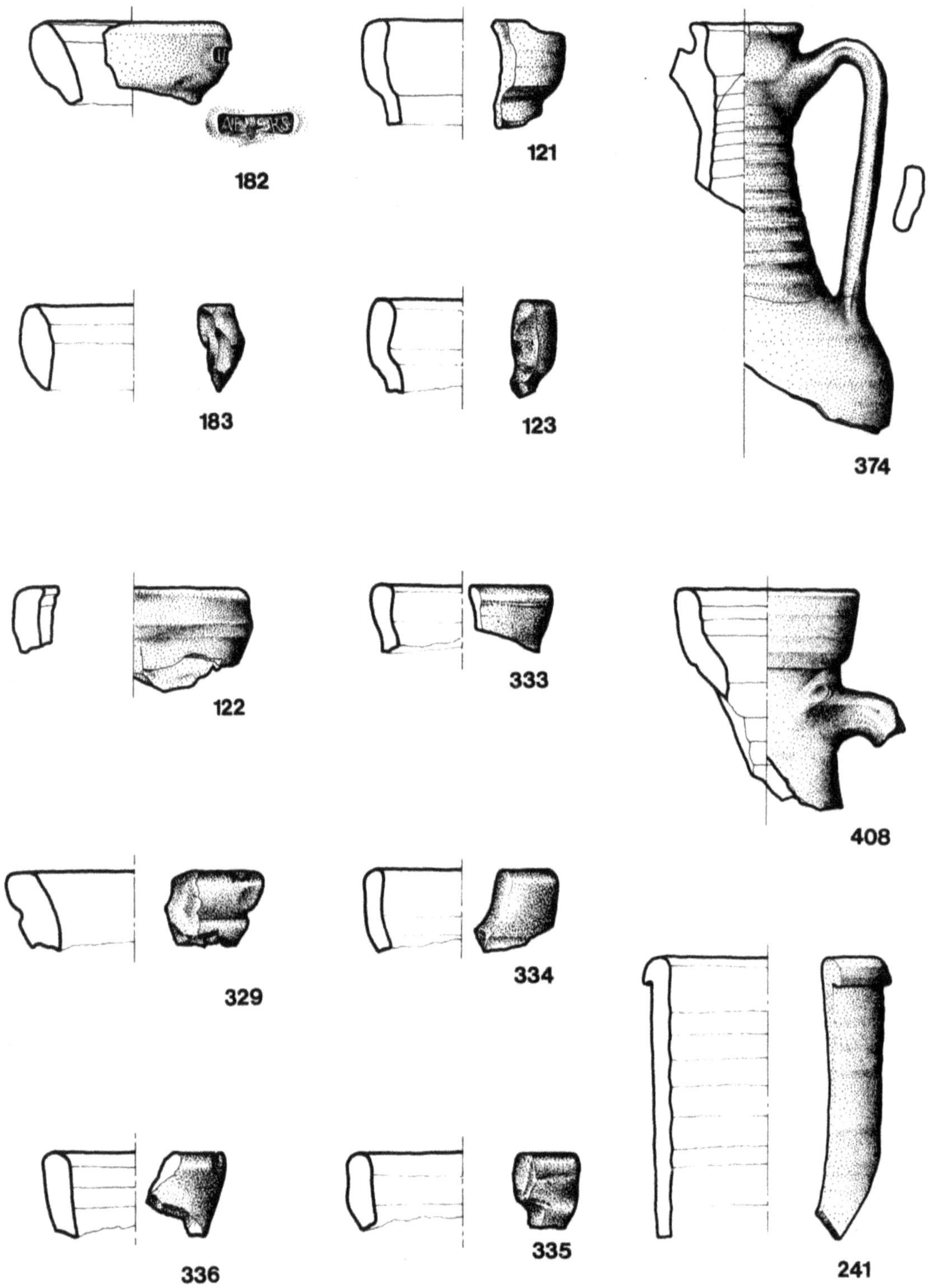

Figure 29 Amphorae (nos. 121-123, 329, 333-336) Salla, (nos. 182, 241, 374) Savaria and (nos. 183, 408) Scarbantia

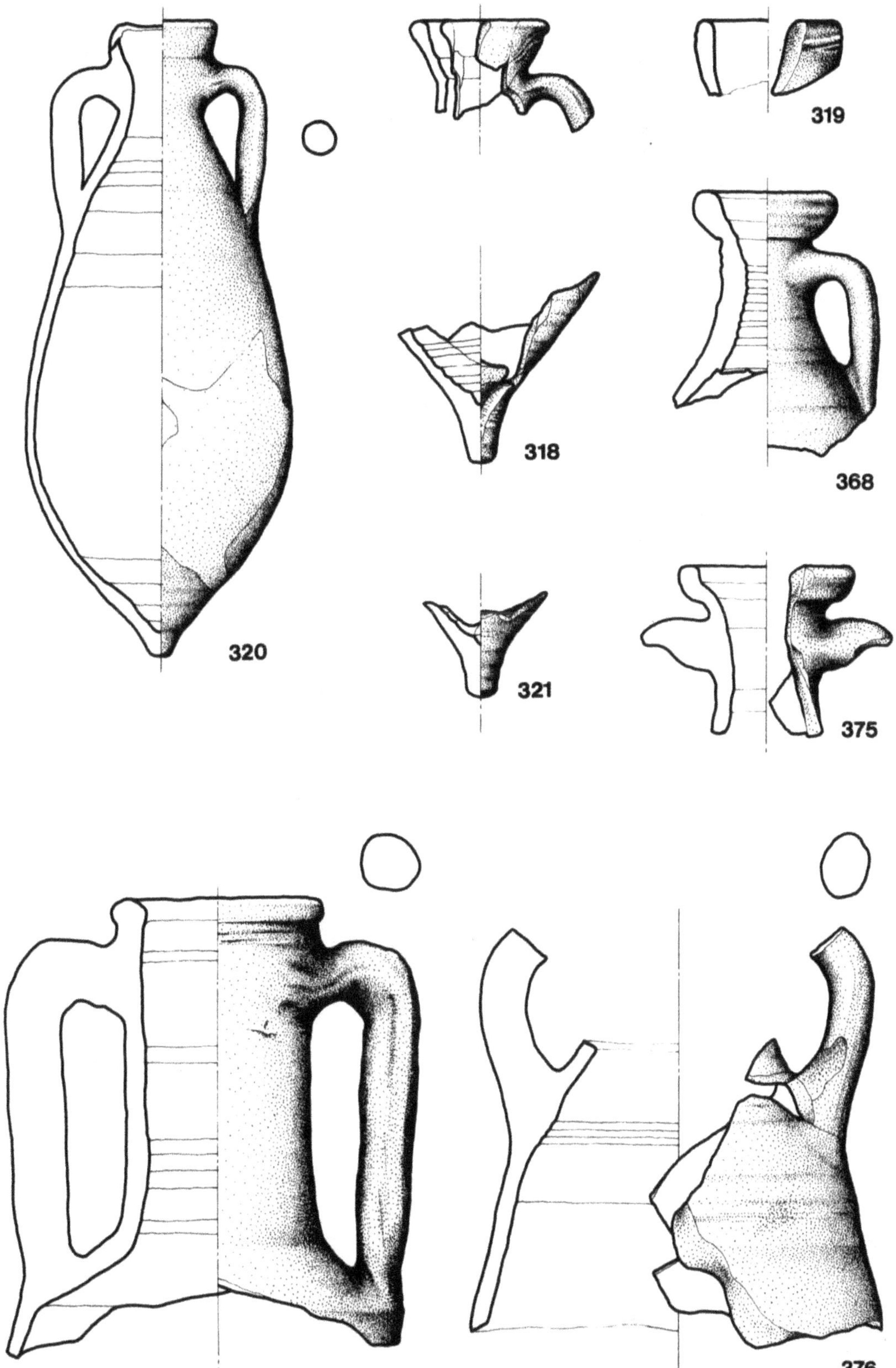

Figure 30 Amphorae (nos. 318–319, 368) Salla, (nos. 320–321, 375–376) Savaria and (no. 409) Scarbantia

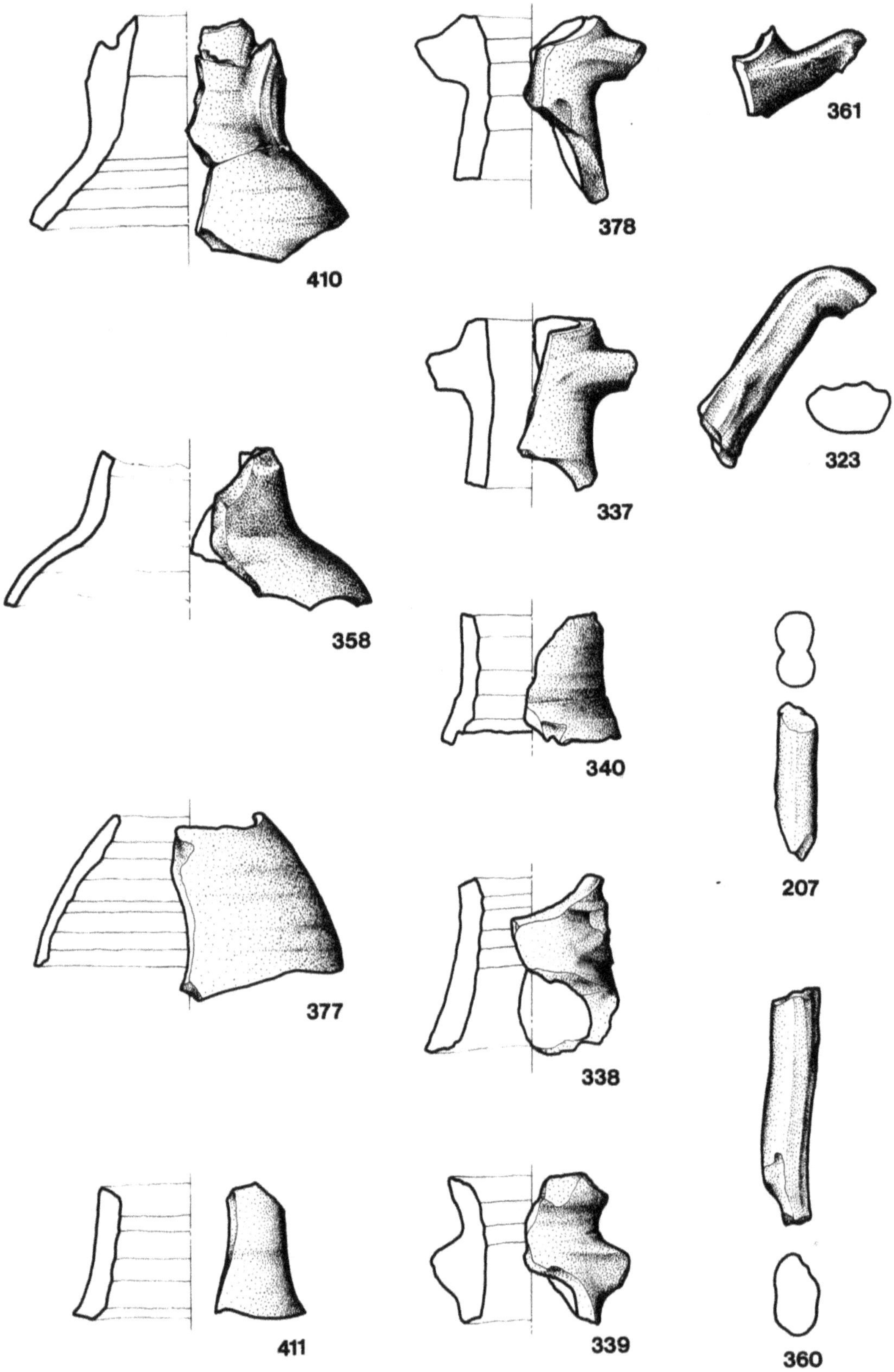

Figure 31 Amphorae (nos. 207, 337-340, 358, 360-361) Salla, (nos. 377-378) Savaria and (nos. 323, 410-411) Scarbantia

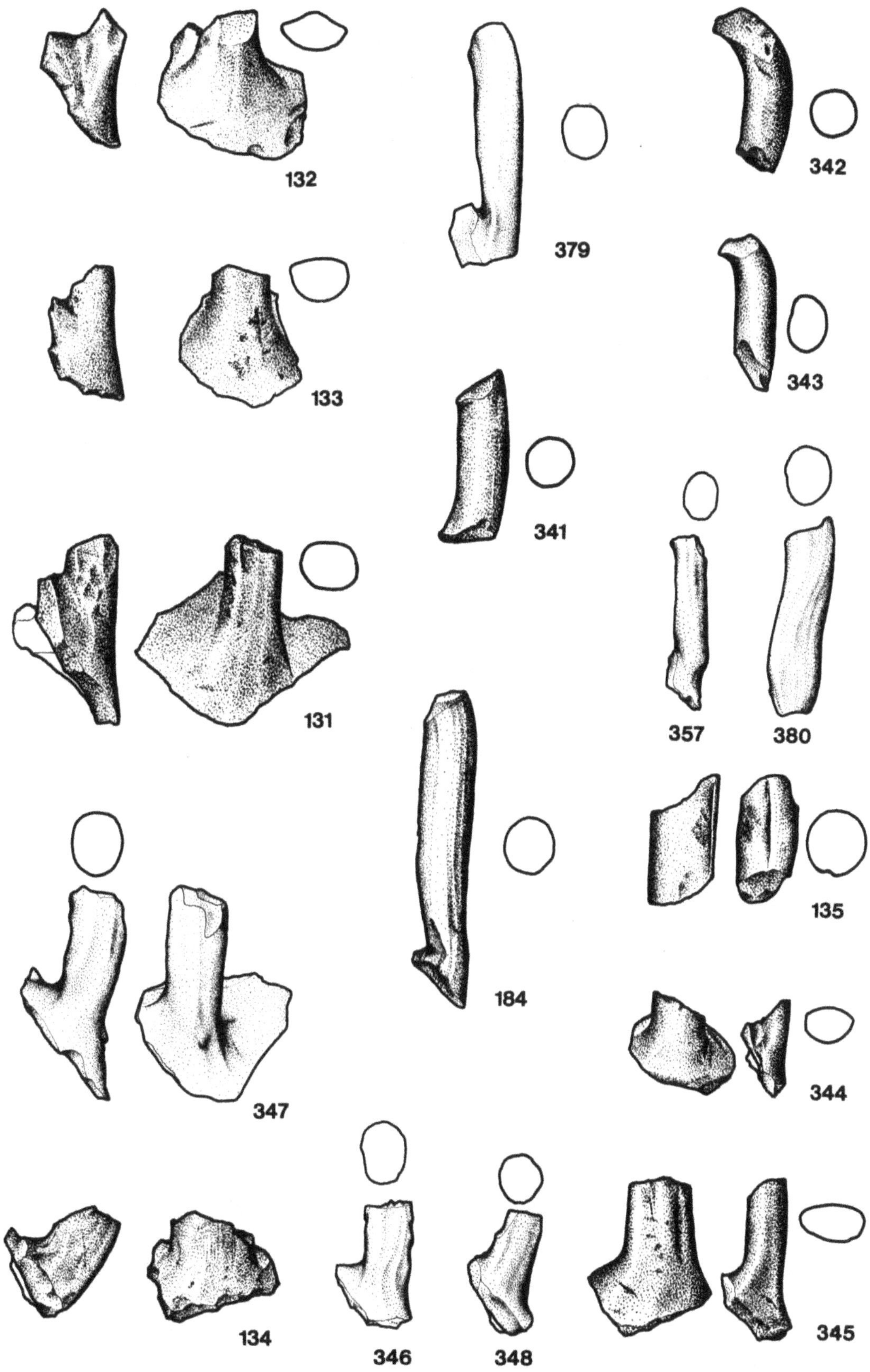

Figure 32 Amphora handles (nos. 131-134, 341-348, 357) Salla, (nos. 379-380) Savaria and (no. 184) Scarbantia

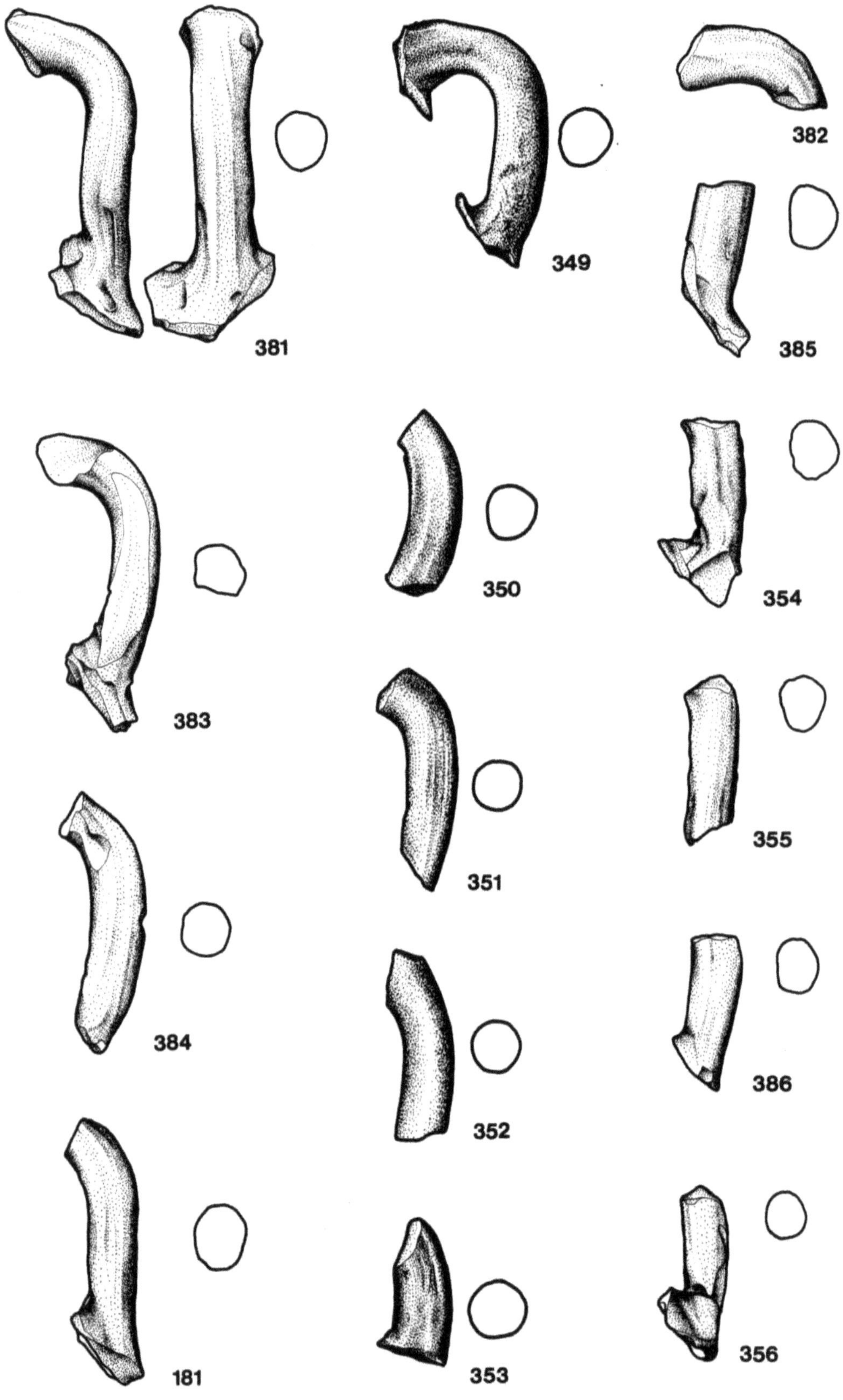

Figure 33 Amphora handles (nos. 349-356) Salla and (nos. 181, 381-386) Savaria

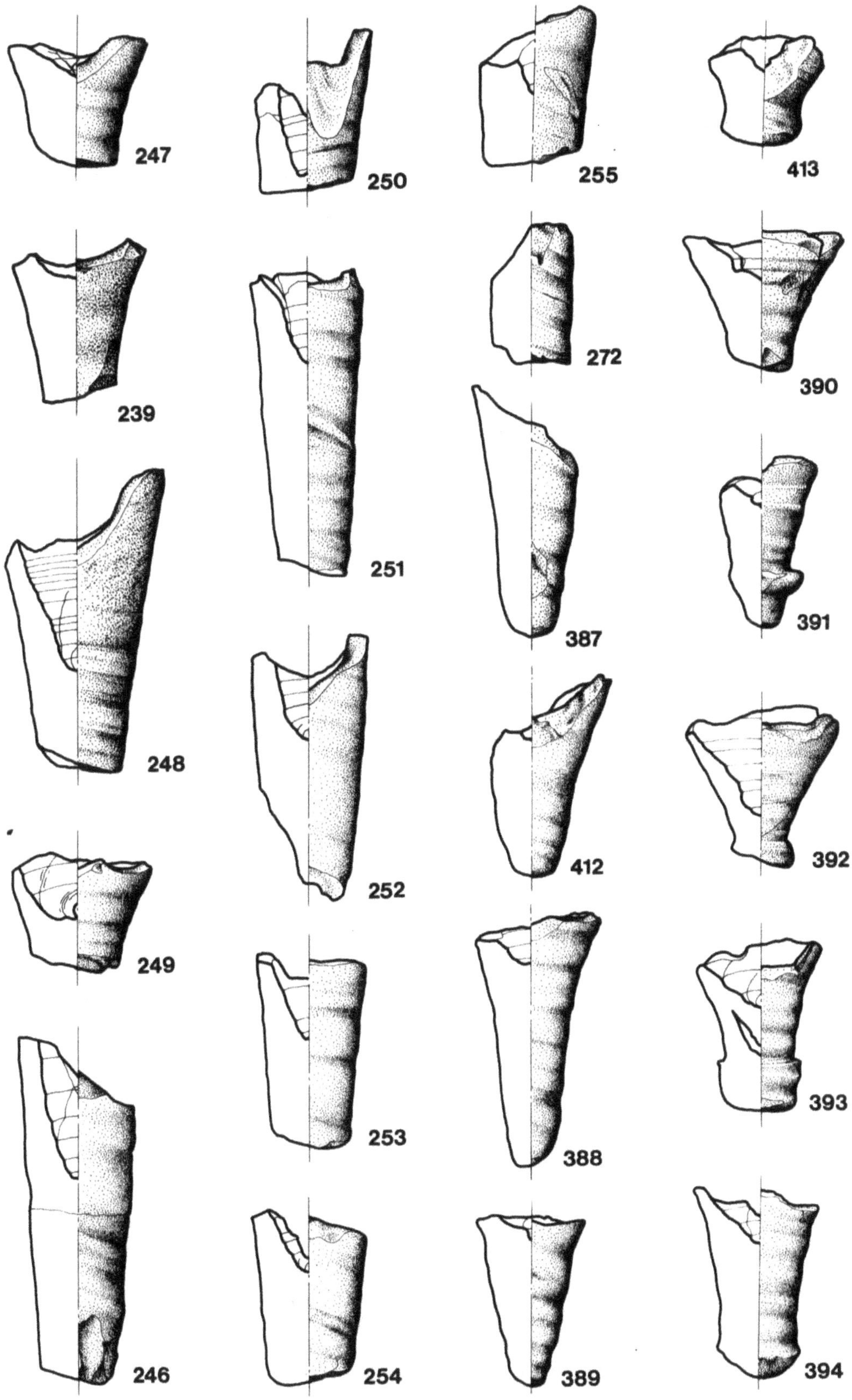

Figure 34 Amphora bases (no. 239) Salla, (nos. 246-255, 387-394) Savaria and (nos. 272, 412-413) Scarbantia

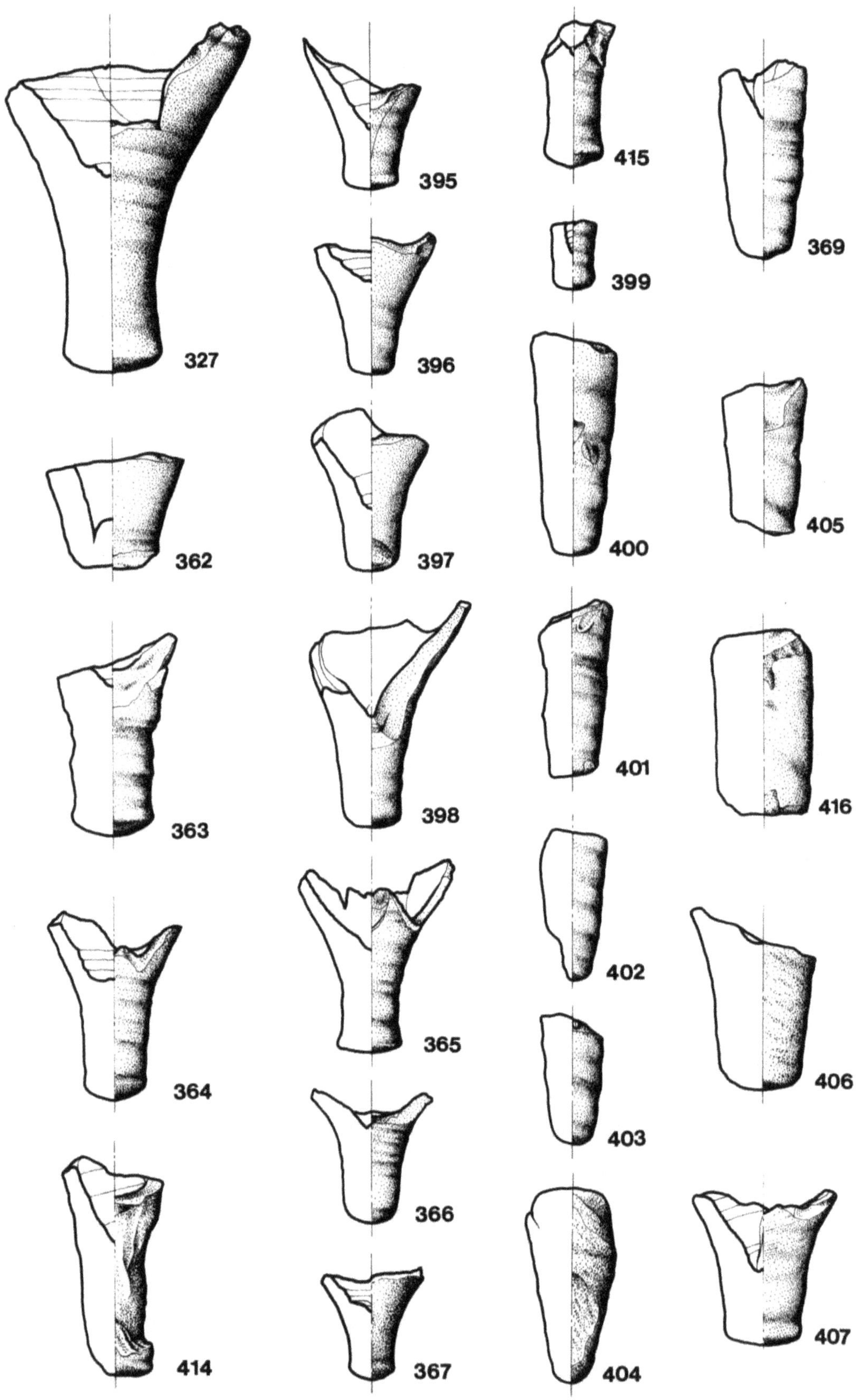

Figure 35 Amphora bases (nos. 362-367, 369) Salla, (nos. 395-407) Savaria and (nos. 327, 414-416) Scarbantia

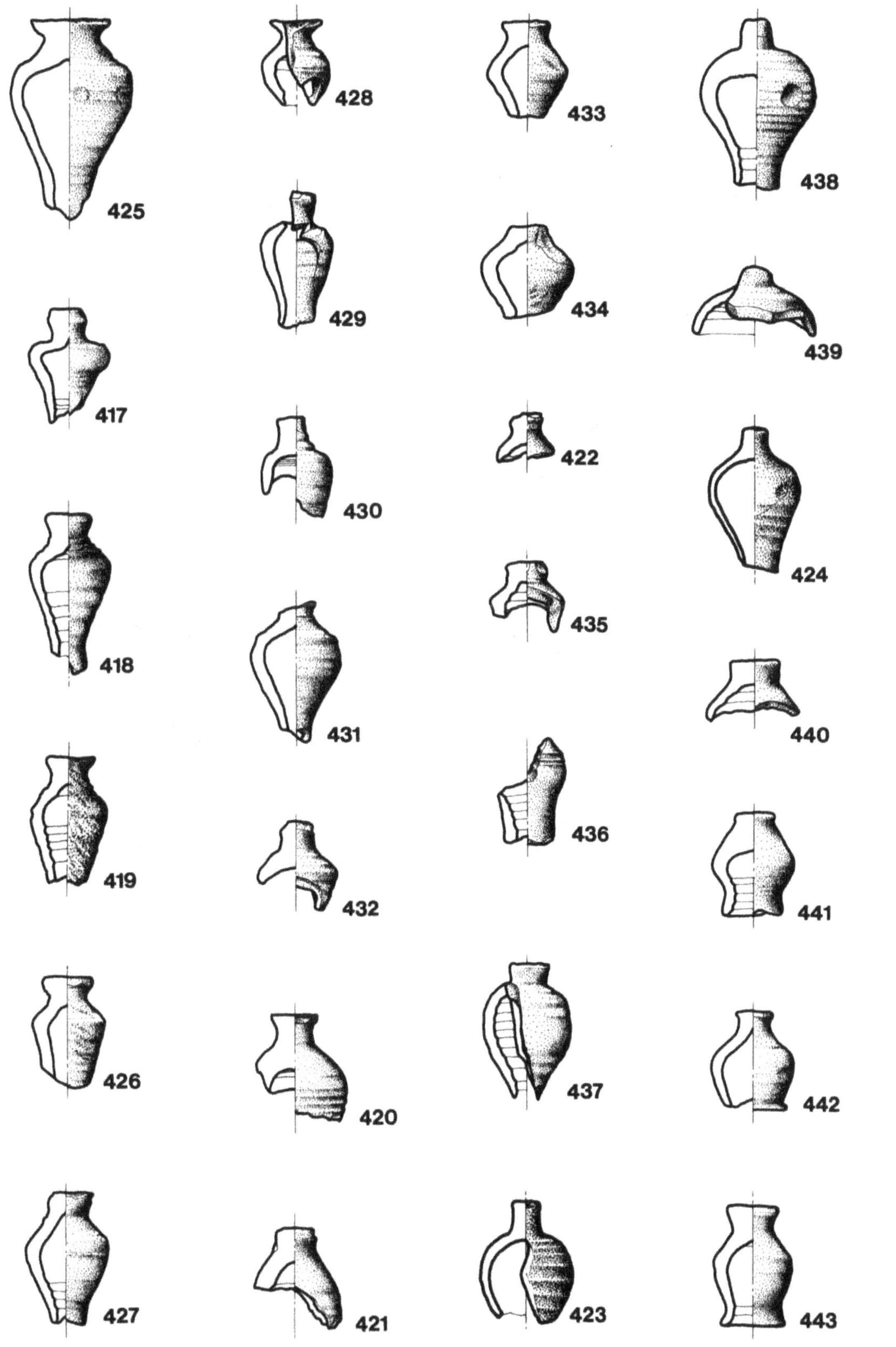

Figure 36 Amphora stoppers (417-424) Salla, (nos. 425-442) Savaria and (no. 443) Scarbantia

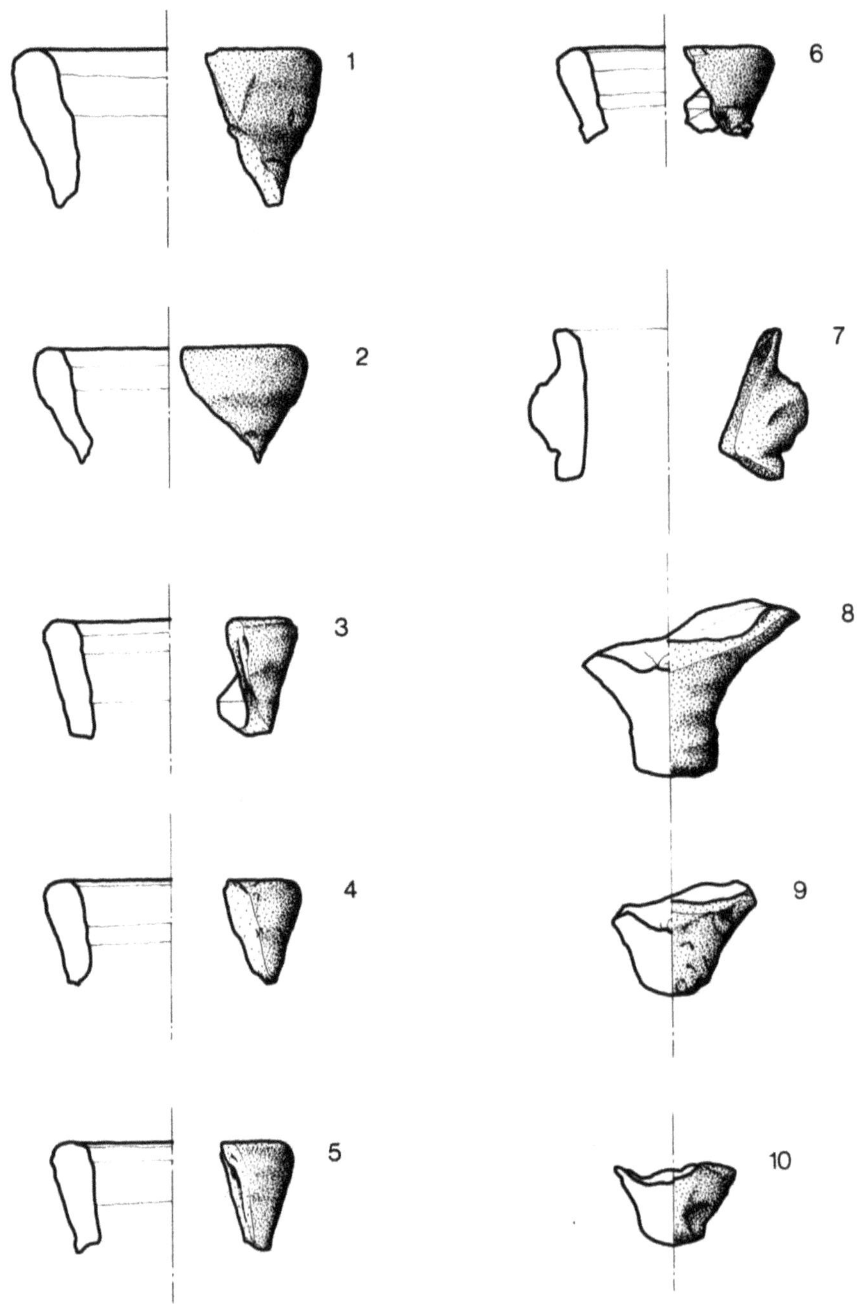

Figure 37 Dressel 6B amphorae from Loron
1 (no. 9), 2 (no. 8), 3 (no. 5), 4 (no. 10), 5 (no. 6),
6 (no. 4), 7 (no. 3), 8 (no. 7), 9 (no. 2), 10 (no. 1)

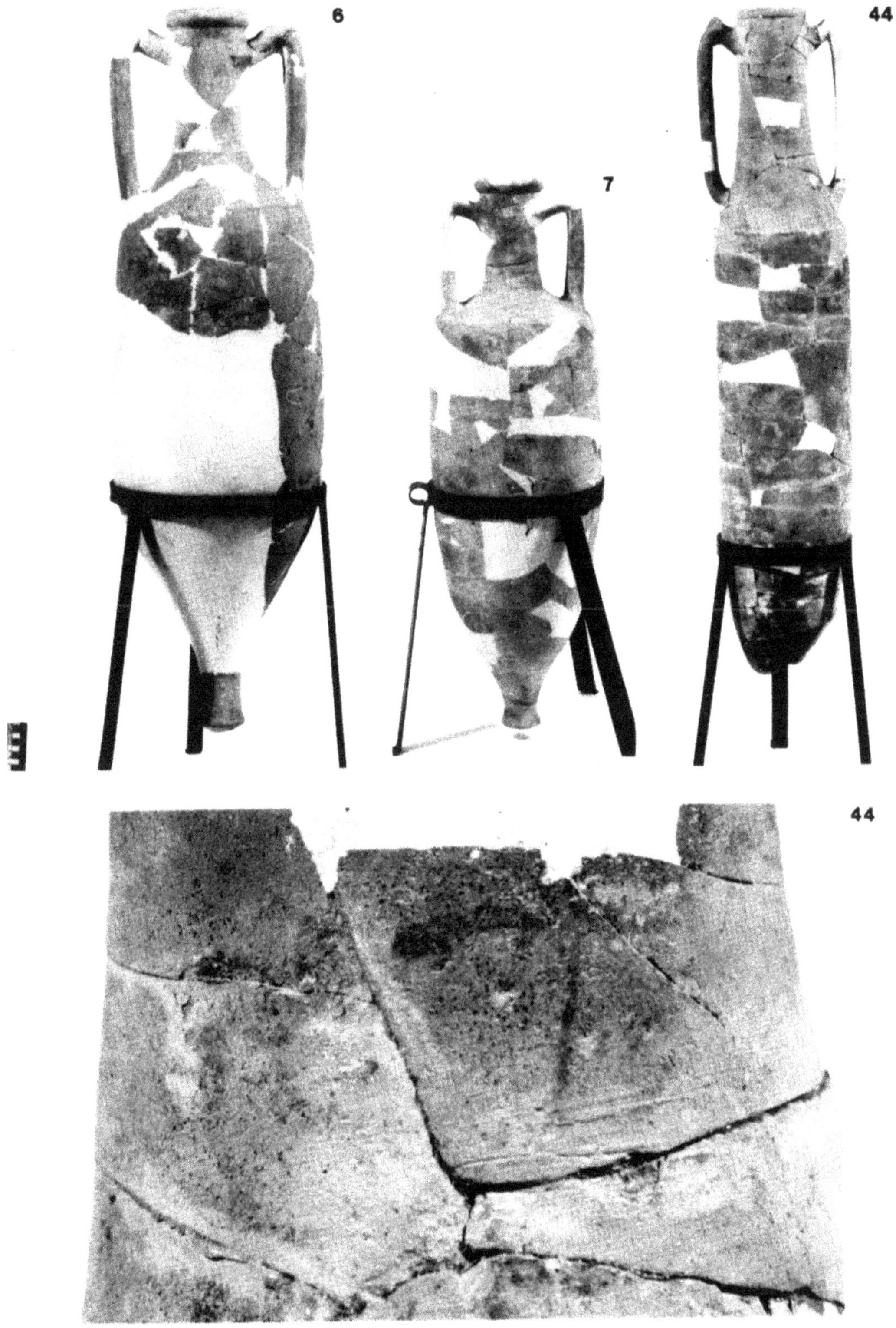

Plate 1 Dressel 2-4 (nos. 6-7), and Dressel 5 (no. 44) amphorae from Poetovio

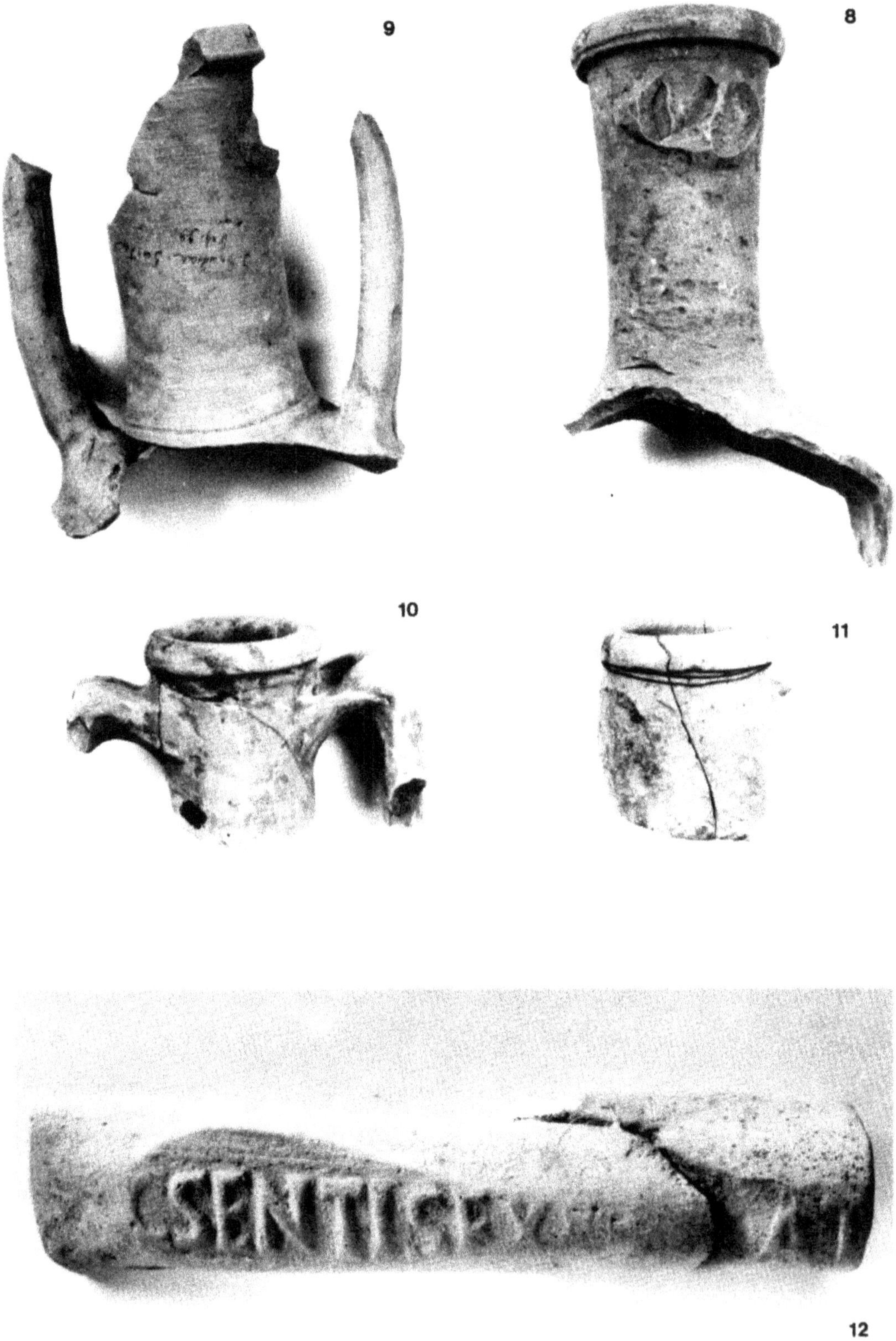

Plate 2 Dressel 2-4 (nos. 8-12) amphorae from Poetovio

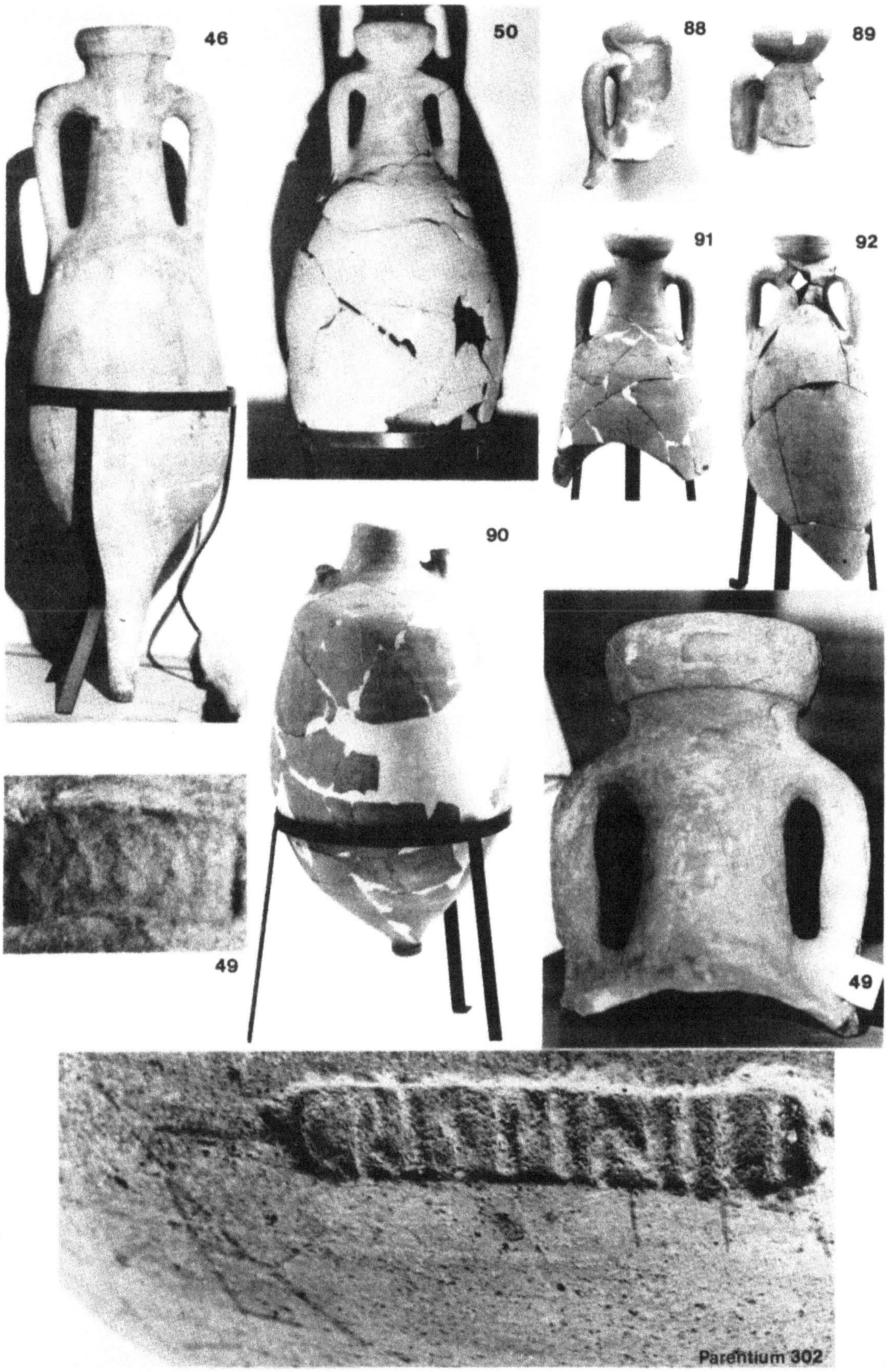

Plate 3 Dressel 6A (no. 46) amphora from Emona, Dressel 6B amphorae (no. 49) from Nauportus (stamp CAMI), (no. 50) from Emona, (nos. 88–92) from Poetovio and CRISPINILI stamp with Laecanius grafitto on a rim (Parentium 302) from Poreč.

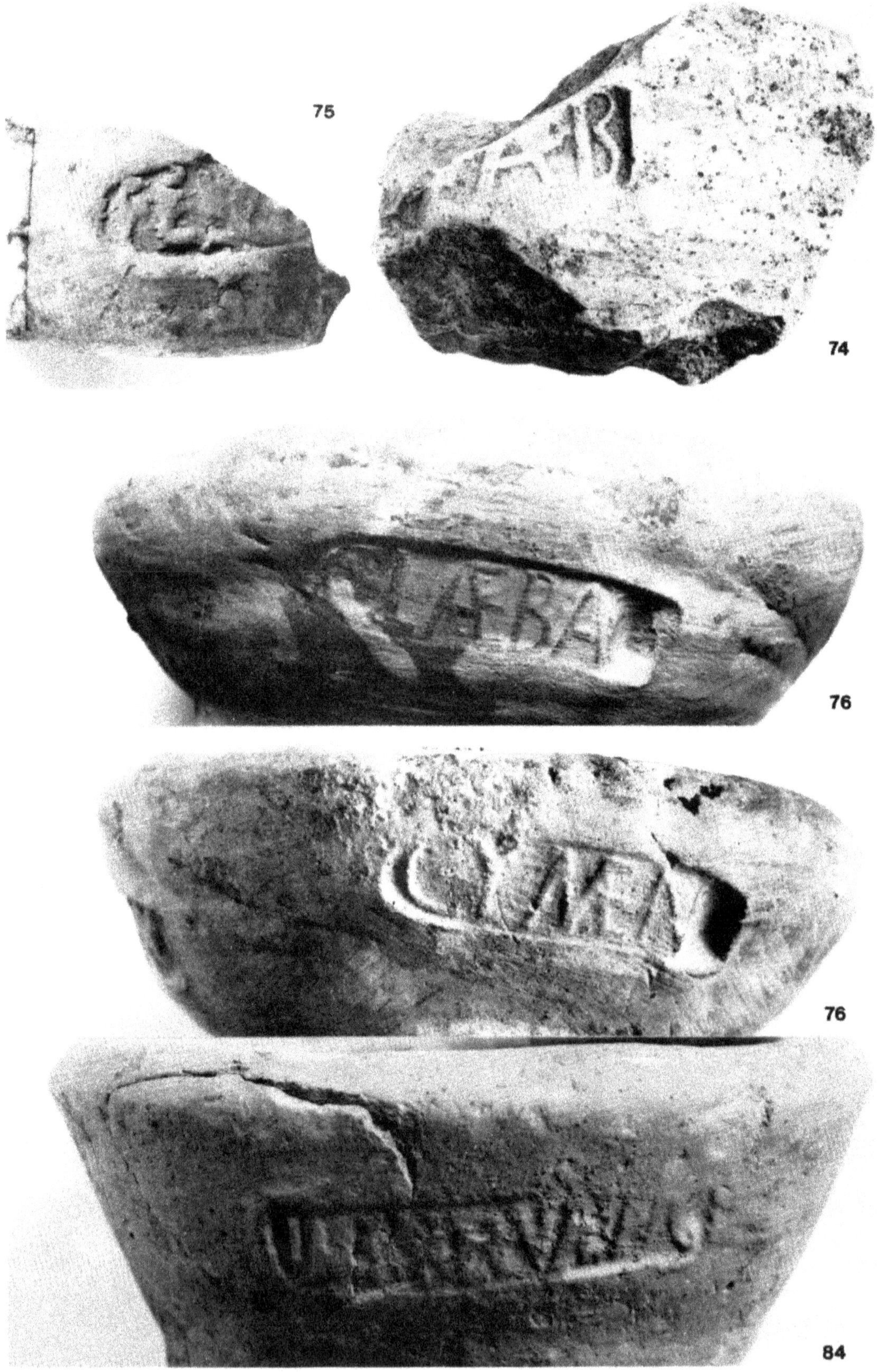

Plate 4 Dressel 6B amphorae from Poetovio (no. 75, stamp FELC), (no. 74, stamp LA$\overline{EK}$B), (no. 76, C.LA$\overline{EK}$.BAS; CL$\overline{Y}$MEN), (no. 84. IMPN$\overline{ER}$VA$\overline{E}$A$\overline{V}$G).

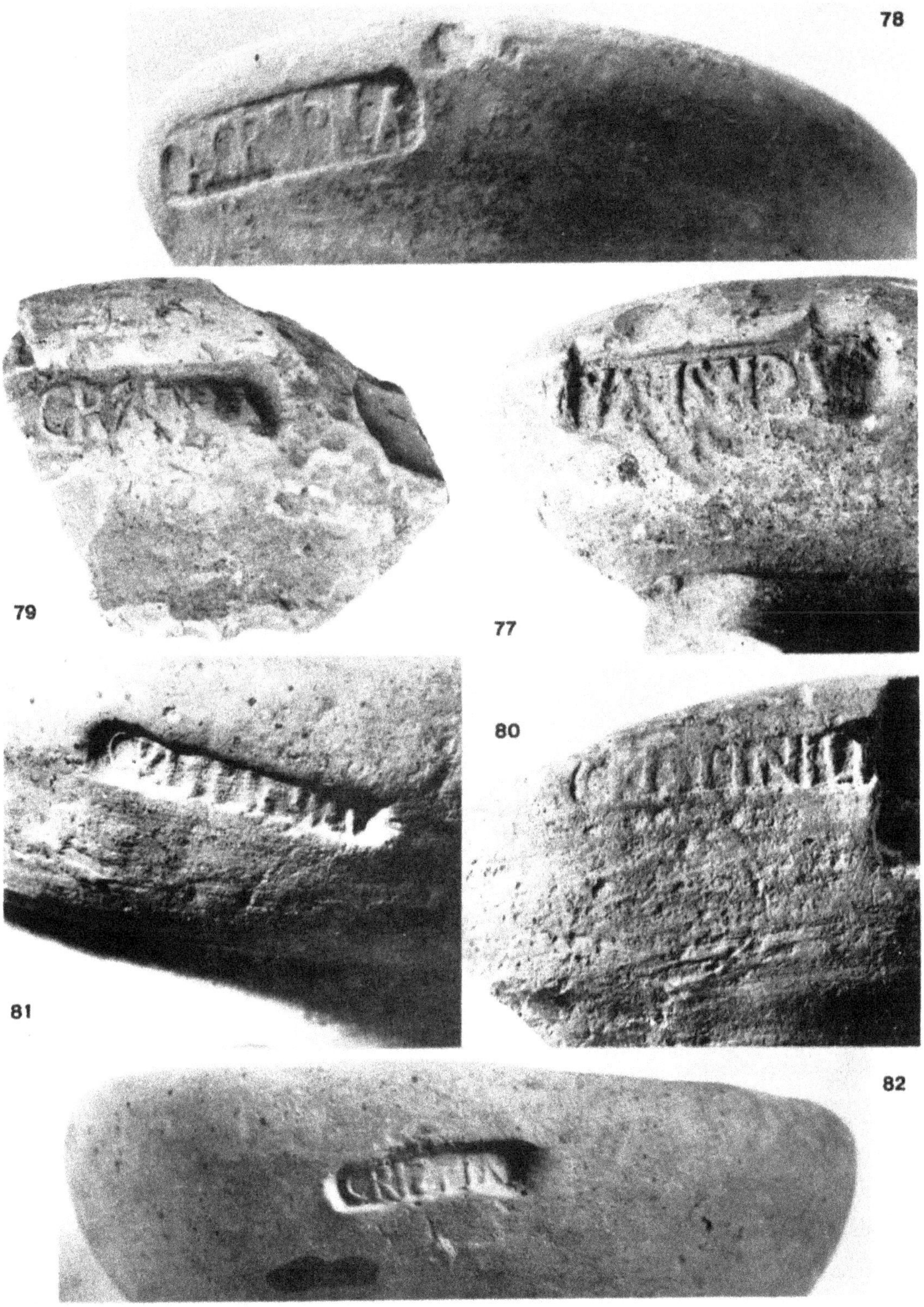

Plate 5 Dressel 6B amphorae from Poetovio (nos. 77–79, stamp C$\overline{\text{AL}}$.C$\overline{\text{RISPINILLAE}}$), (nos. 80–81, stamp CRISPINILI), (no. 82, stamp CRIƧPIN).

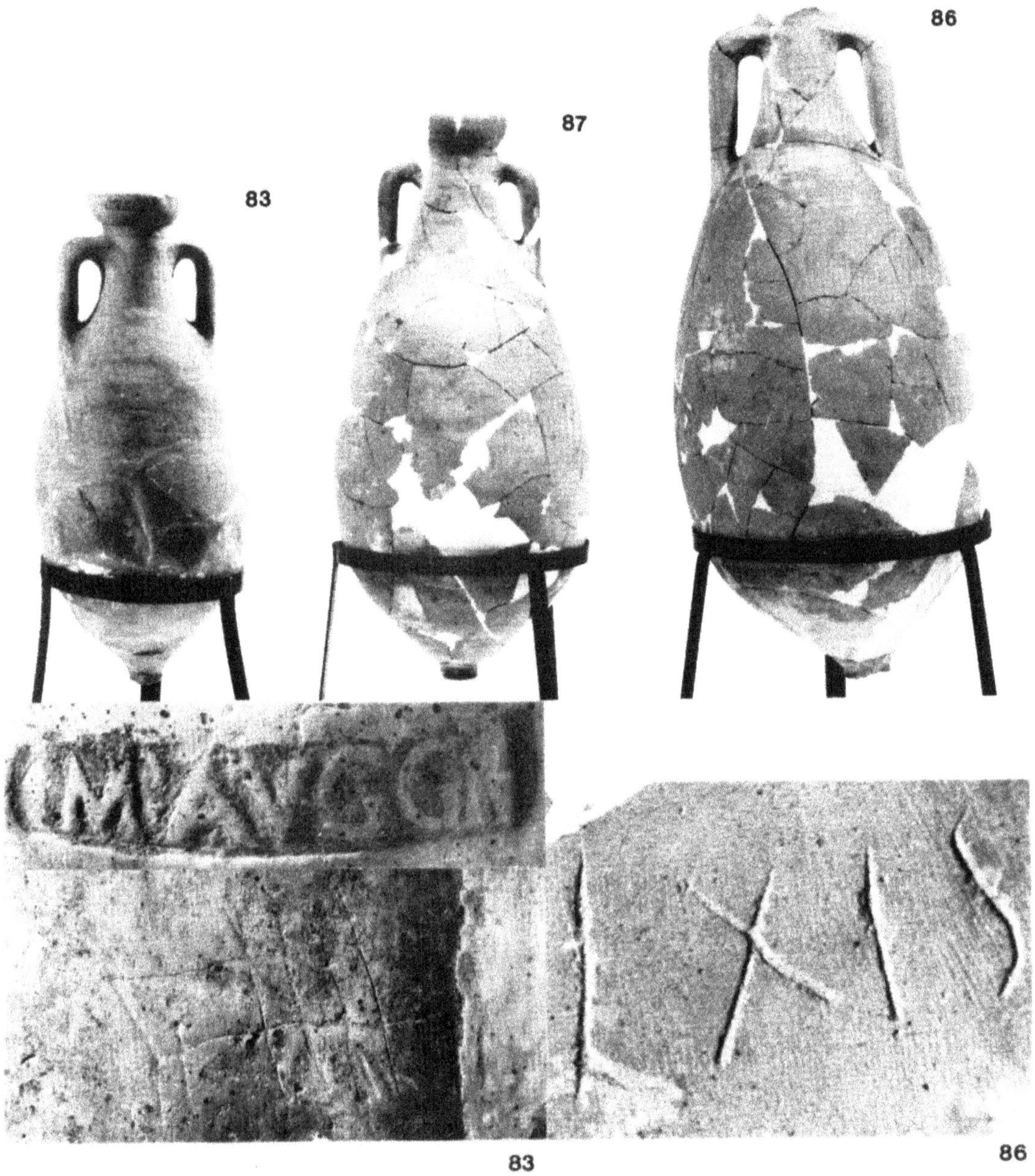

Plate 6 Dressel 6B amphora from Poetovio (no. 83, stamp I$\overline{MP}$.AVG.G$\overline{ER}$).

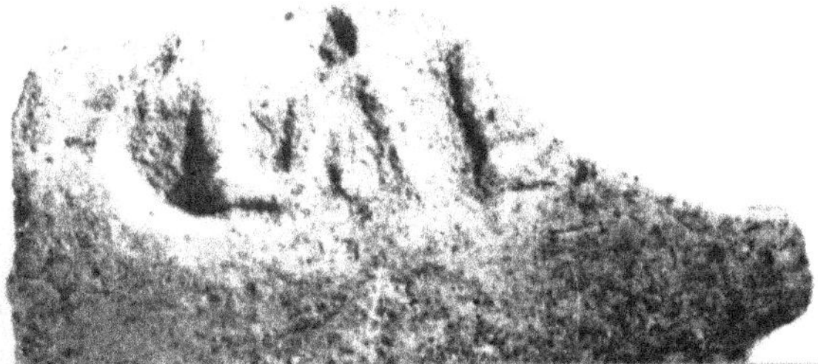

96

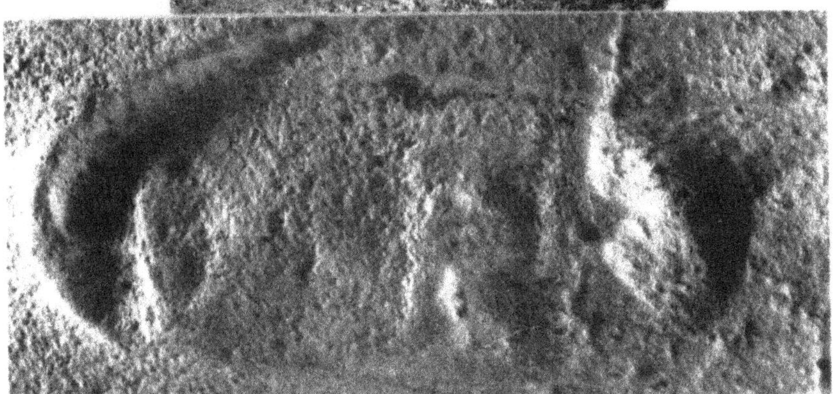

150

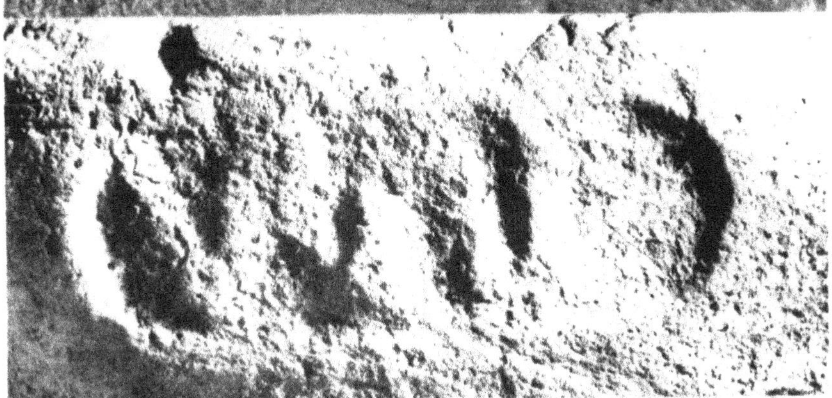

150

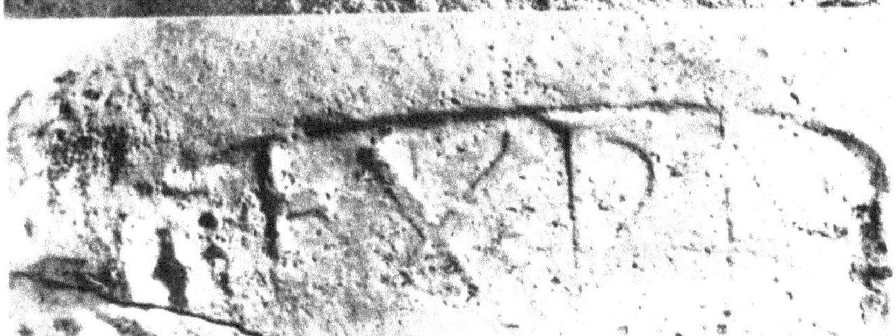

97

107

Plate 7 Dressel 6B amphorae (no. 96, stamp LAE, no. 97, stamp FEX.PE, no. 107, FEL.T$\overline{VR}$) from Salla and (no. 150, $\overline{IA}$Ek, VIAT) Savaria.

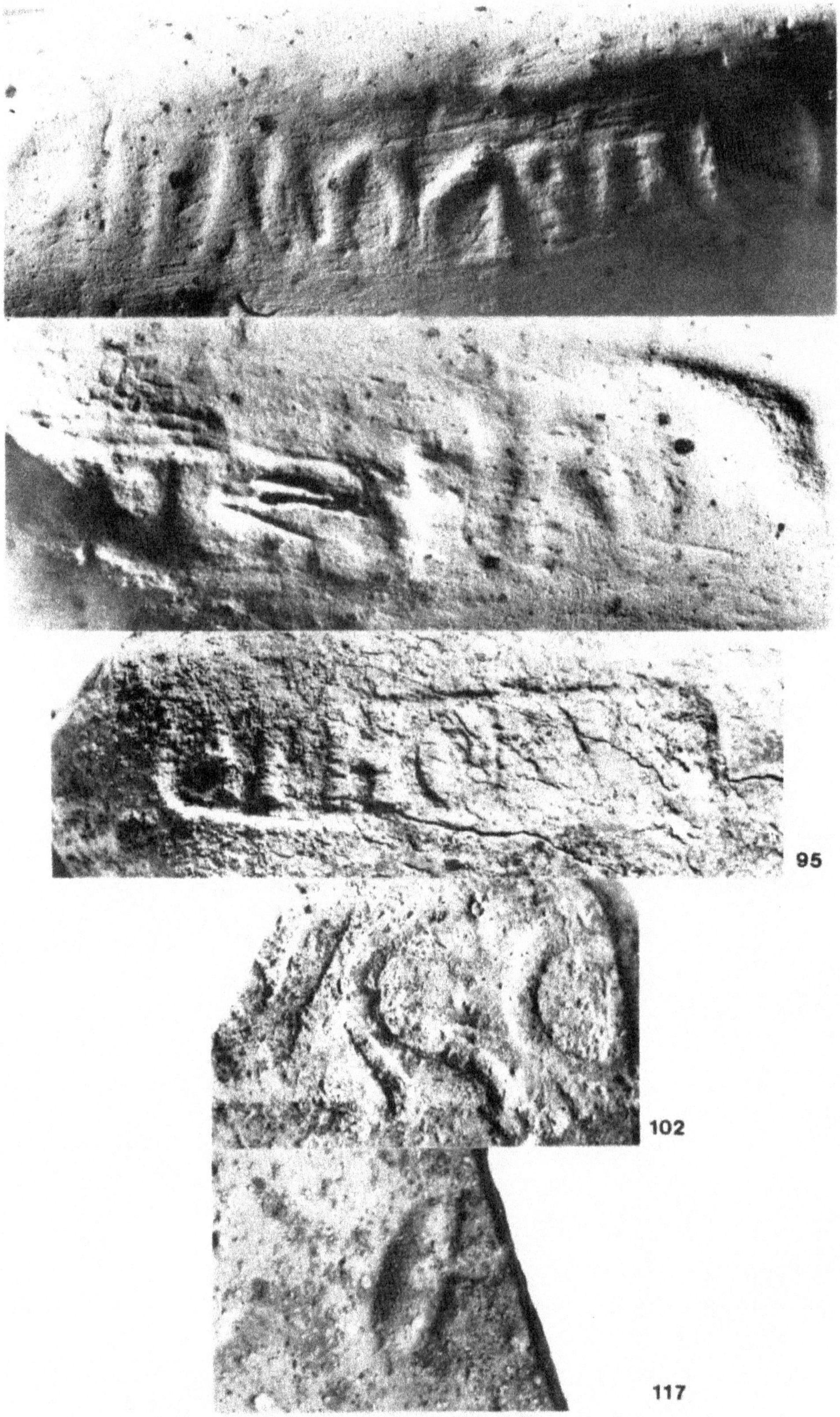

154

154

95

102

117

Plate 8 Dressel 6B amphorae (no. 95, stamp APHORVS, no. 102, stamp ..AVGG.., no. 117, stamp A..) from Salla and (no. 154, stamps ITIMOCꟼMI, .ESR) from Savaria.

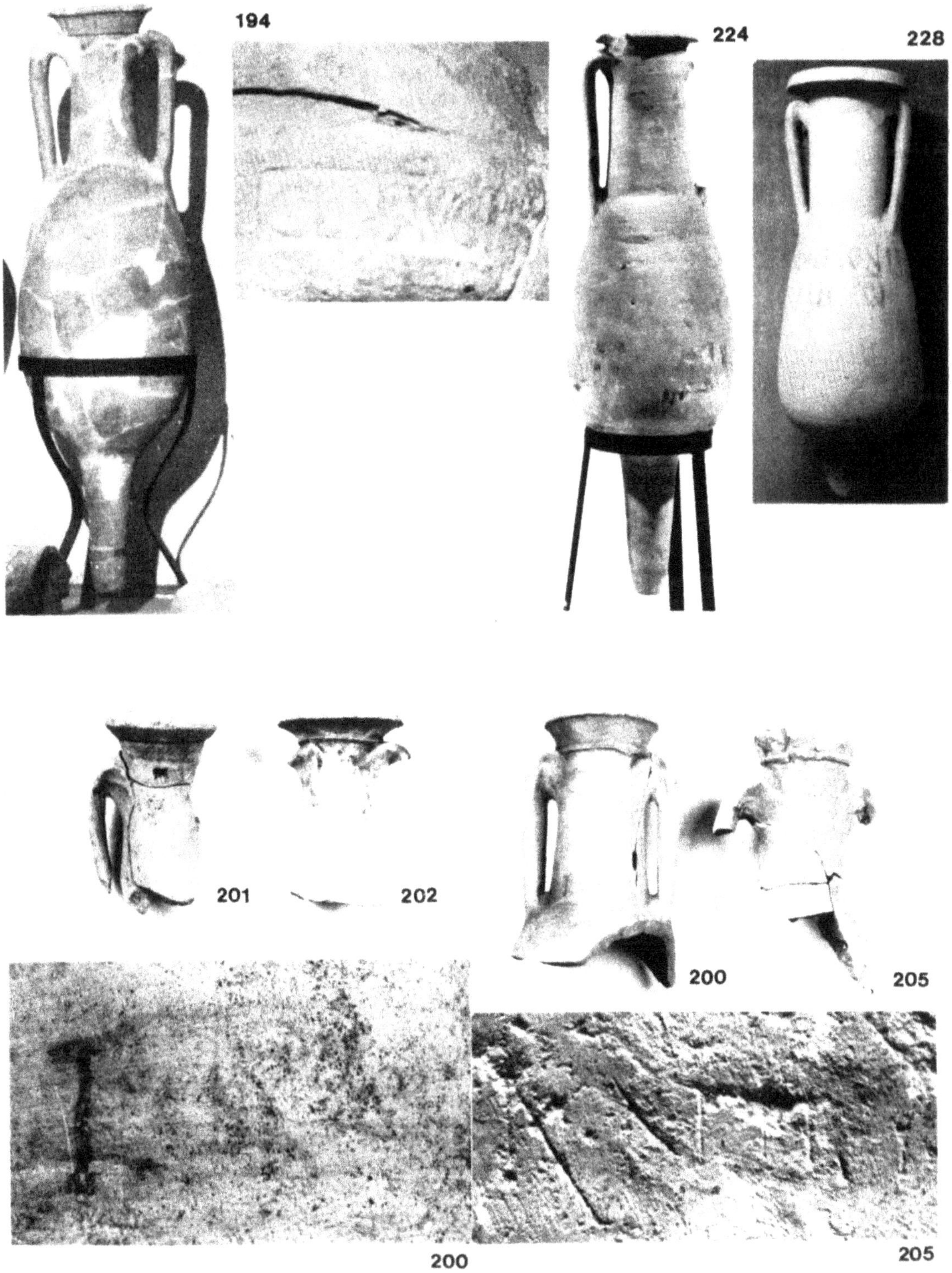

Plate 9 Dressel 7-11 amphorae (no. 194, stamp CL$\overline{ANI}$.AVG$\overline{TI}$) from Emona, (no. 200, red painted inscription T.$\overline{TR}$.A; nos. 201-202, no. 205. grafitto K.AIIII) from Poetovio, Pélichet 46 amphorae (no. 224) from Poetovio and (no. 228, red painted inscription) from Carnuntum.

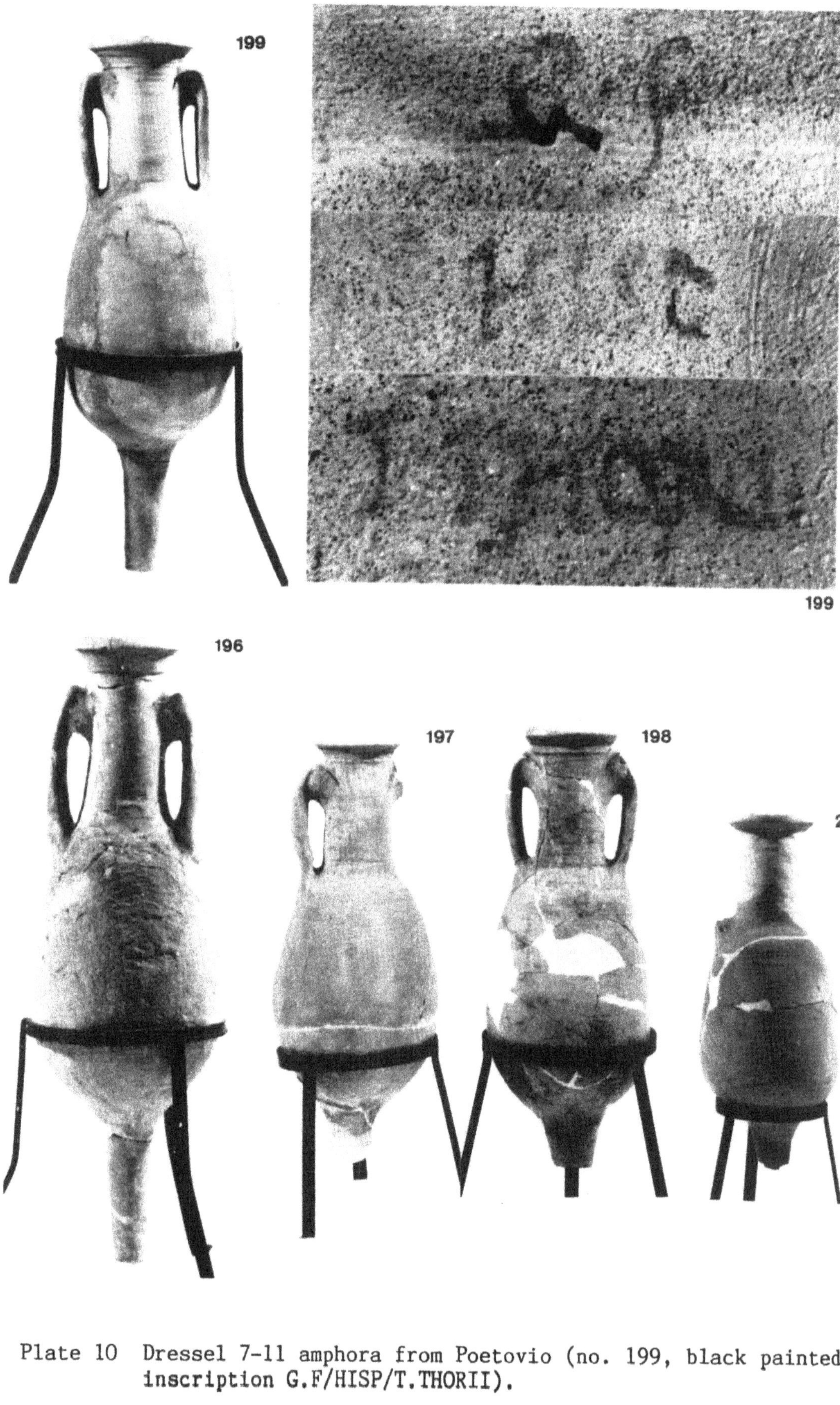

Plate 10 Dressel 7-11 amphora from Poetovio (no. 199, black painted inscription G.F/HISP/T.THORII).

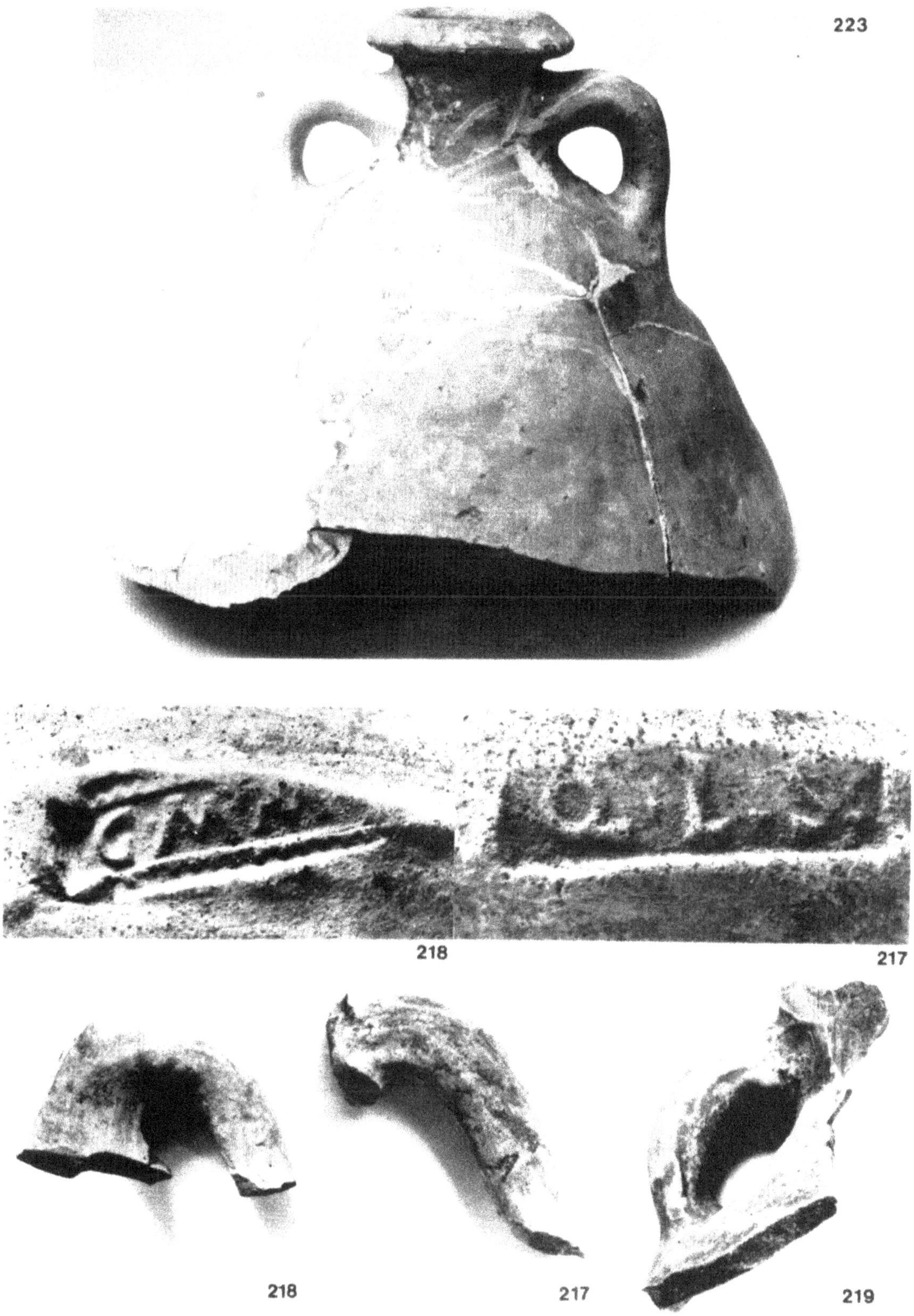

Plate 11 Dressel 20 amphorae (no. 223) from Savaria and (no. 217, stamp Q.L.S.; no. 218, stamp GMM; no. 219) from Poetovio

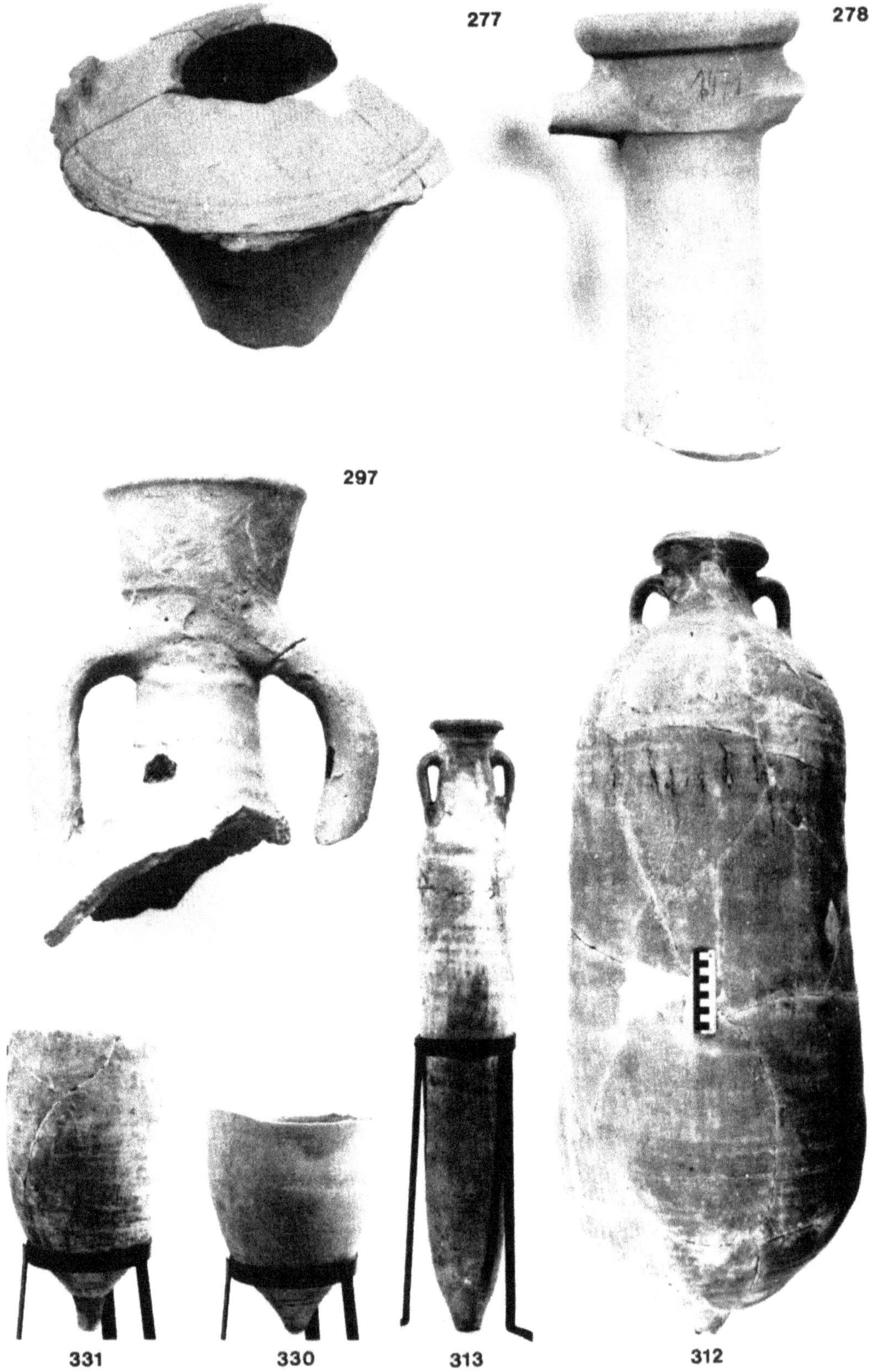

Plate 12 Schörgendorfer 558 amphorae (nos. 277-278), Portorecanation amphora (no. 297), African (no. 312) and Spateion (no. 313) amphorae, from Poetovio.

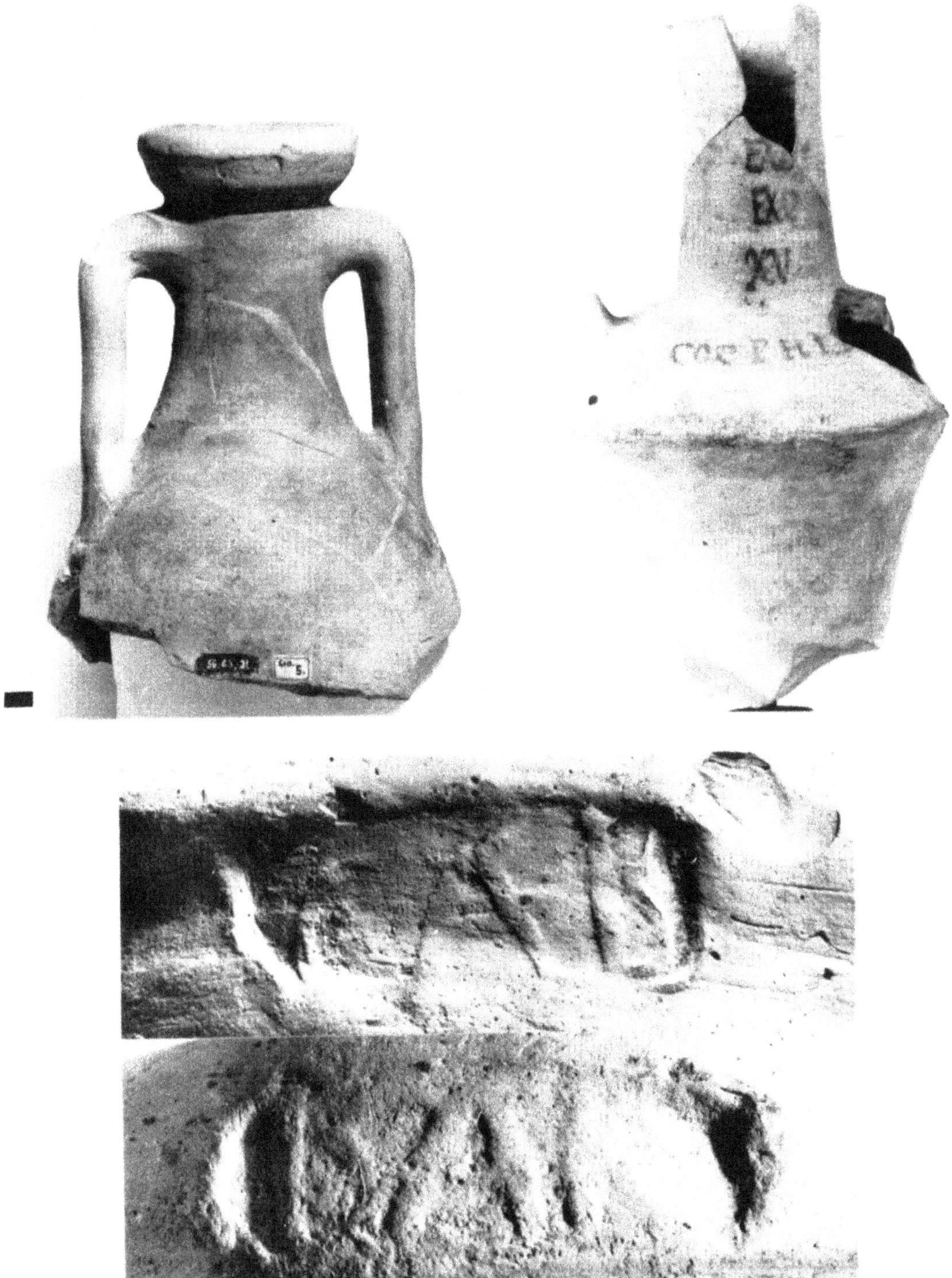

Plate 13 Laecanius stamps (CL$\overline{\text{AEKB}}$, BAR) on Dressel 6B and Schörgendorfer 558 (red painted inscription EXD/EXC/XV/COC PHIL) amphora from Aquincum.

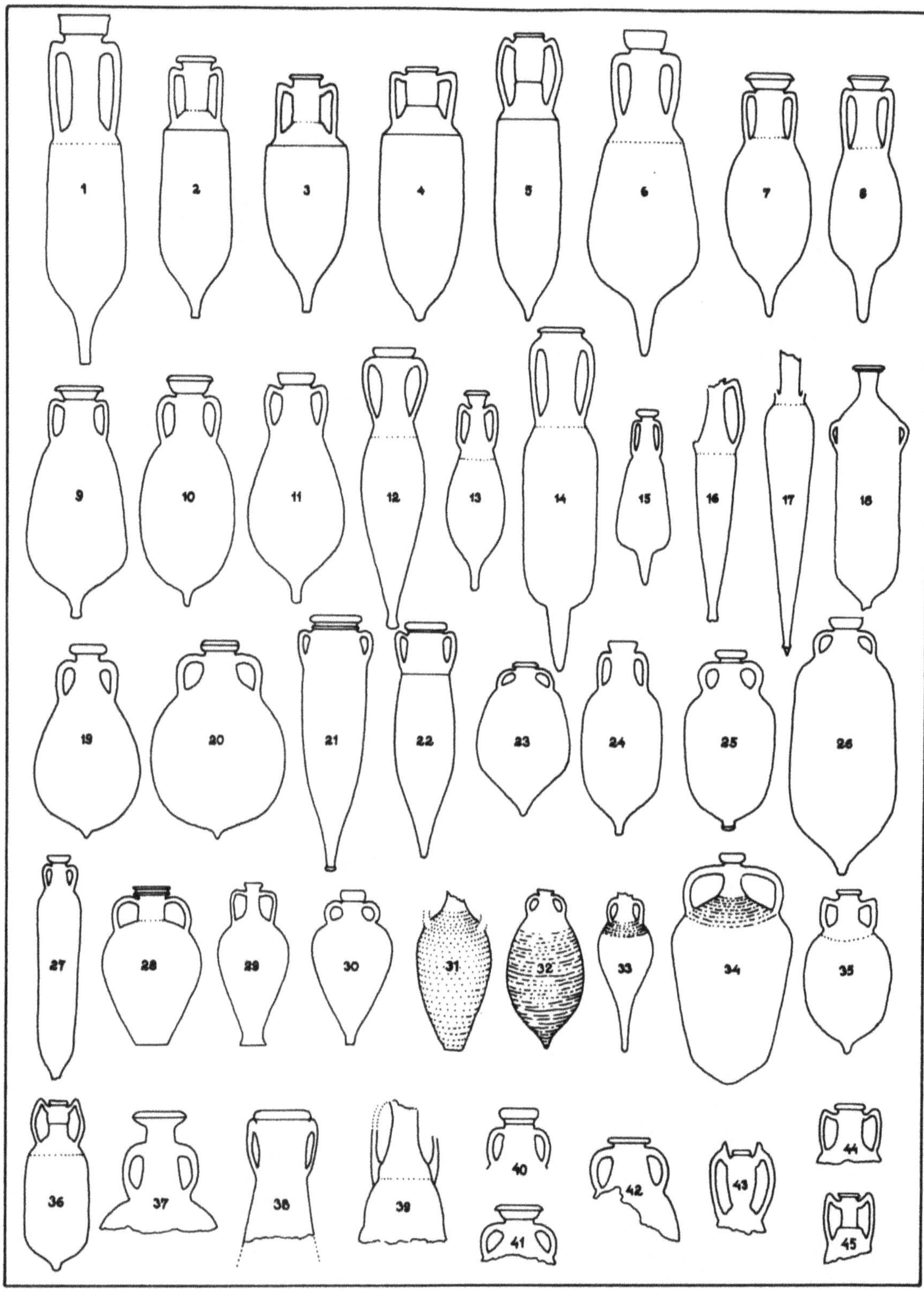

Corp. inscr. Lat. vol. XV.

AMPHORARVM FORMAE

Figure 38. The Dressel table of amphora forms

Supplement 2

Tamás Bezeczky, Pál Kerékfy & Lajos Rónyai

An application of computer statistics for the classification of Roman amphorae

We tackled the problems posed by the classification of these vessels in the research on Roman amphorae we undertook in the period 1982-85.

Our aim was to find geometrical invariants which would make classification more accurate. As our starting point, we took the methodology of A. Guénoche and A. Tchernia[160] and both extended and modified it on the basis of the experience gathered. The most important outcome of our work is a more reliable and a more accurate mode of description, which allows us to perform data processing and analysis by computer. The techniques utilized (primarily ideas from computer statistics) can be applied to other artefacts as well.

The problem

In research related to amphorae, an important part is played by classification on the basis of form, together with experience gleaned from the study of stamps, inscriptions and the fabric of the vessels. The first significant classification was undertaken and published at the end of the last century by H. Dressel in Bull. Comm. Arch. Roma 1879 and in C.I.L. XV.2.

He was the first to draw the attention of students to the relationship between epigraphic data and forms. He drew up a table of illustrations for the inscriptions and stamps from the Castro Pretorio of Rome and Monte Testaccio at Rome which has been used by archaeologists until the present (fig. 38). This table of typologies, the first and undertaken with exemplary thoroughness, is still the most comprehensive published. Despite apparent deficiencies, archaeologists are constrained to use it because no other comprehensive work has been able to replace it. Many suggestions have been made as to how the table could be supplemented.[161] In the majority of cases such suggestions are aimed at an exhaustive treatment of the individual types or groups of types. A wide range of activities must be assessed because amphorae were made for the transport and storage of so many different foodstuffs (oil, olives, wine, fish sauce), from various provinces of the Roman world. The problem is further complicated by the fact that amphorae of the same or similar types could be made in different places. The objective of the present work is to place these typological investigations, so far undertaken primarily on an experimental basis, on precise, quantified grounds. That is to say, we

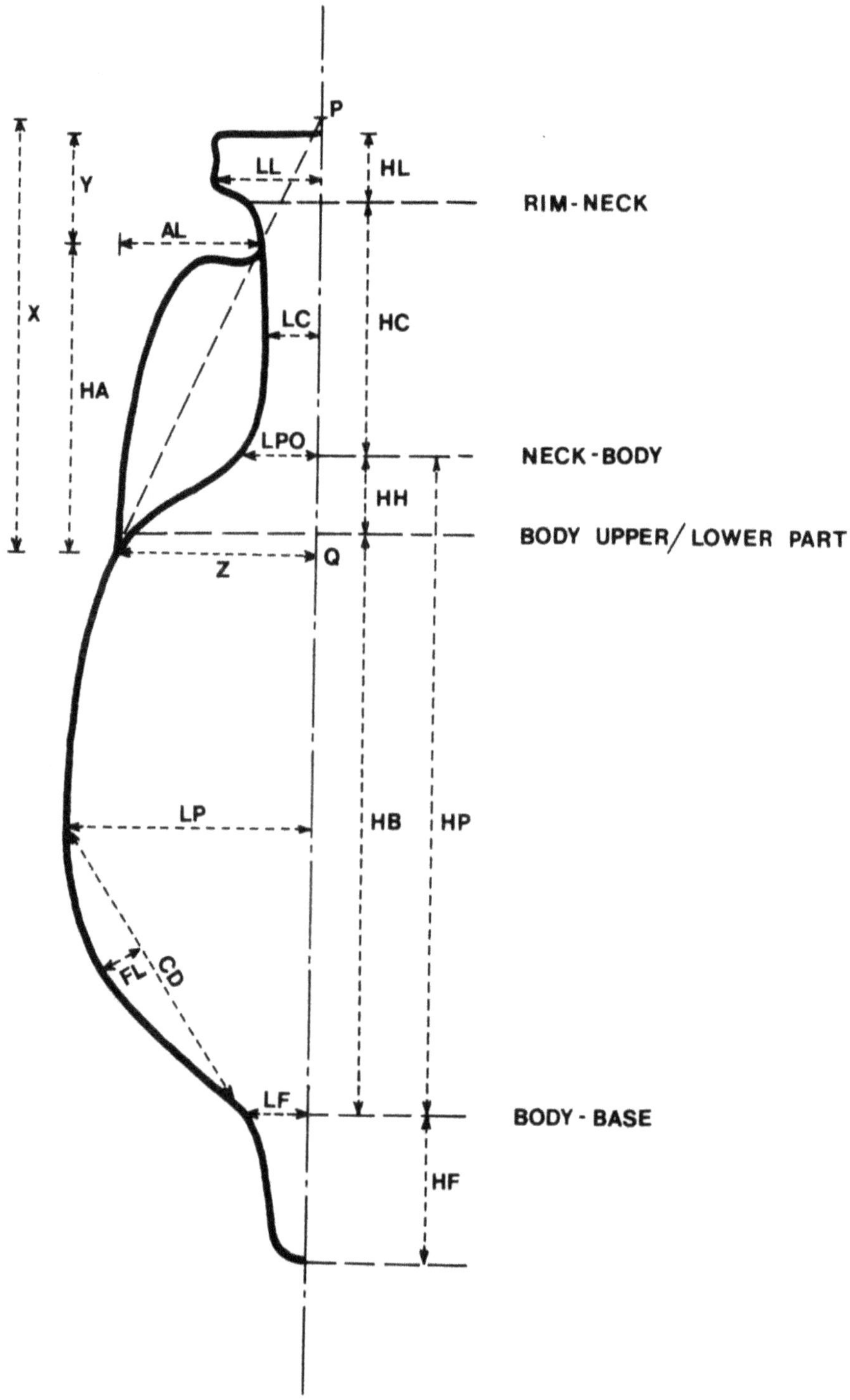

Figure 39. Segments of the amphora

tried to introduce such measurable well-defined attributes which grasp the information hidden in the geometry and shape of the amphorae as accurately and reliably as possible.

The Guénoche and Tchernia method

As a starting point, we used the descriptive method of the above researchers, based on the principles enunciated by Hamon and Hesnard.

The values adopted are the following (fig. 39).

LL - the maximum radius of the rim

HL - the height of the mouth

LC - the minimum radius of the neck

HC - the height of the neck

LPO - the radius at the junction of the body and neck

LP - the maximum radius of the body

HH - the height of the upper part of the body

HB - the height of the lower part of the body

HT - the complete height of the amphorae

LF - the radius at the junction of the body and the base

HF - the height of the base

AL - the width of the handle

HA - the height of the handle

CD - the length of the line segment connecting the junction of the body and the base to the point where the maximum diameter of the body of the amphora is attained

FL - the maximum distance between the lower part of the body and the CD line segment

Z - the distance between the lower part of the handle and the longitudinal axis

X - let P denote the common point of the straight line connecting the joining points of the handles and of the longitudinal axis, and let Q be the axial point of the line segment Z. The distance X is the line segment connecting the points P and Q i.e. its length

Y - the difference between the upper level of the amphora and the junction with the handle.

From the above, the following ratios are set up:

HT= HL + HC + HH + HB + HF

HP= HH + HB

X(25)= HL / (HL + HC)

X(26)= HC / HP

X(27)= LL / HL

X(28)= LC / HC

X(29)= (LL - LC) / LL

X(30)= LP / HT

X(31)= LP / HP

X(32)= HB / HP

X(33)= FL / CD

X(34)= LPO / LP

X(35)= HH / HP

X(36)= LF / LPO

X(37)= HF / HT

X(38)= LF / HF

X(39)= AL / HA

X(40)= Z / X

X(41)= Y / (HL + HC)

Cluster analysis

In order to make the following clear, it is necessary to describe briefly the concept of cluster analysis.

Our basic task is to classify pottery. This means that we want to define groups (clusters) of objects which are homogeneous in the sense that their elements are "similar". Similarity is generally difficult to define, being a subjective concept. It is necessary, therefore to set up a model in which similarity and difference between individual amphorae is a precisely defined notion which may be measured effectively. Such is, for example, a distance to be calculated from results of measurements. On the other hand, this number is expected to express some kind of similarity, proximity etc.

The professional computer statistics packages (e.g. the BMDP we used) offer a variety of such methods.[162]

The initial studies

Amphorae of different types were selected for the investigations, and the geometrical sizes were taken from drawings of individual vessels. Drawings were taken from Panella 1970. Ostia II. (figs. 40-42).

CASE No.	Label	Ostia II.
1	G001	1
2	G002	2
3	G004	8
4	G005	9
5	G006	10
6	G008	12
7	G009	14
8	G011	16
9	G012	17
10	G013	18
11	G014	19
12	G015	20
13	G016	21
14	G018	23
15	G019	24
16	G020	25
17	G021	26
18	G022	27
19	G023	28
20	G024	29
21	G025	30
22	G026	31
23	G030	46
24	G031	47
25	G032	48
26	G033	49
27	G034	50
28	G035	51
29	G037	53
30	G040	56
31	G041	57
32	G042	58
33	G043	59
34	G046	70
35	G047	71
36	G048	72
37	G053	79
38	G054	85

CASE No.	Label	Ostia II.
39	G055	81
40	G056	82
41	G057	83
42	G058	84
43	G059	85
44	G061	89
45	G063	96
46	G068	111
47	G069	112
48	G070	113
49	G073	116

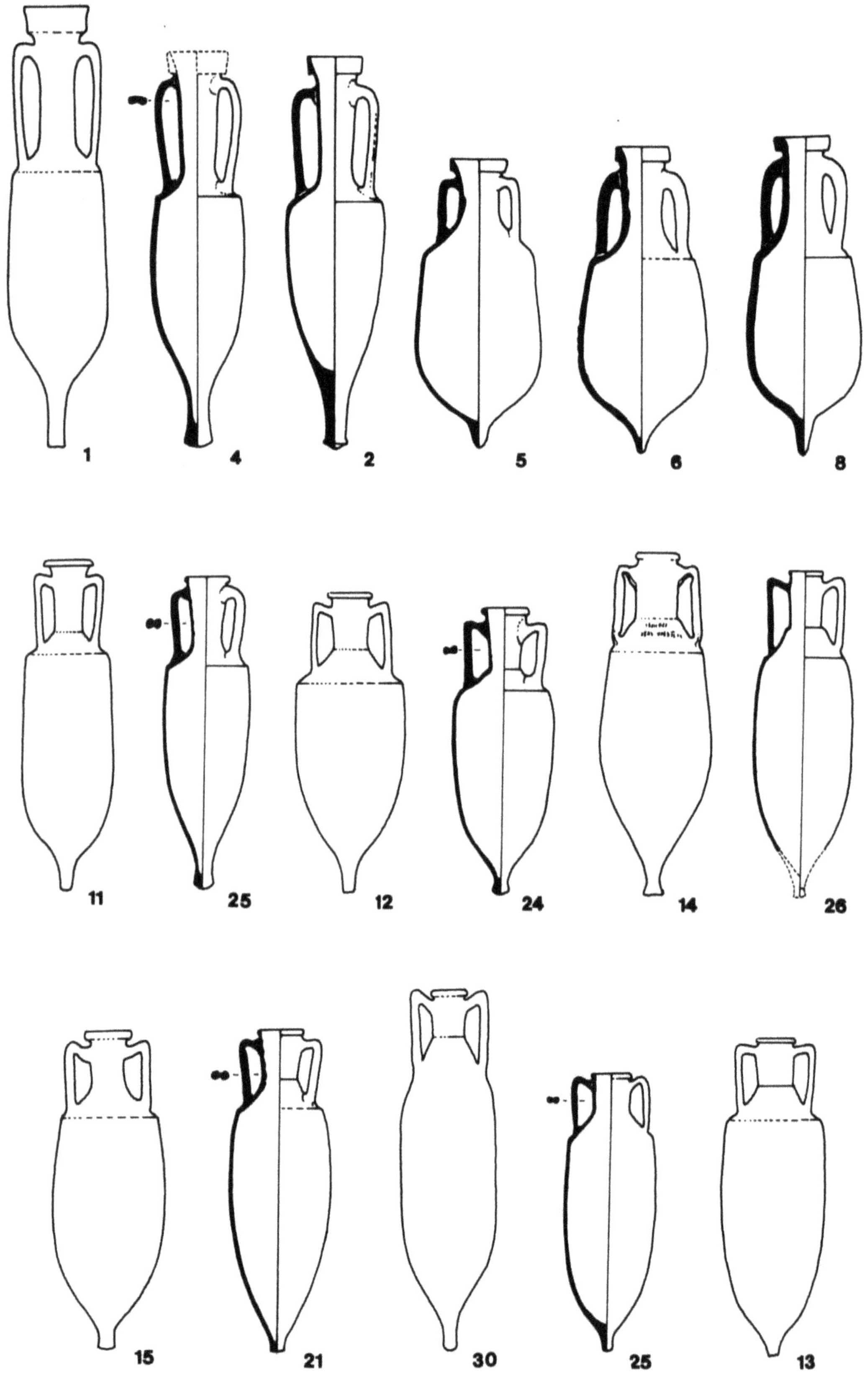

Figure 40. Types of amphorae according to C. Panella

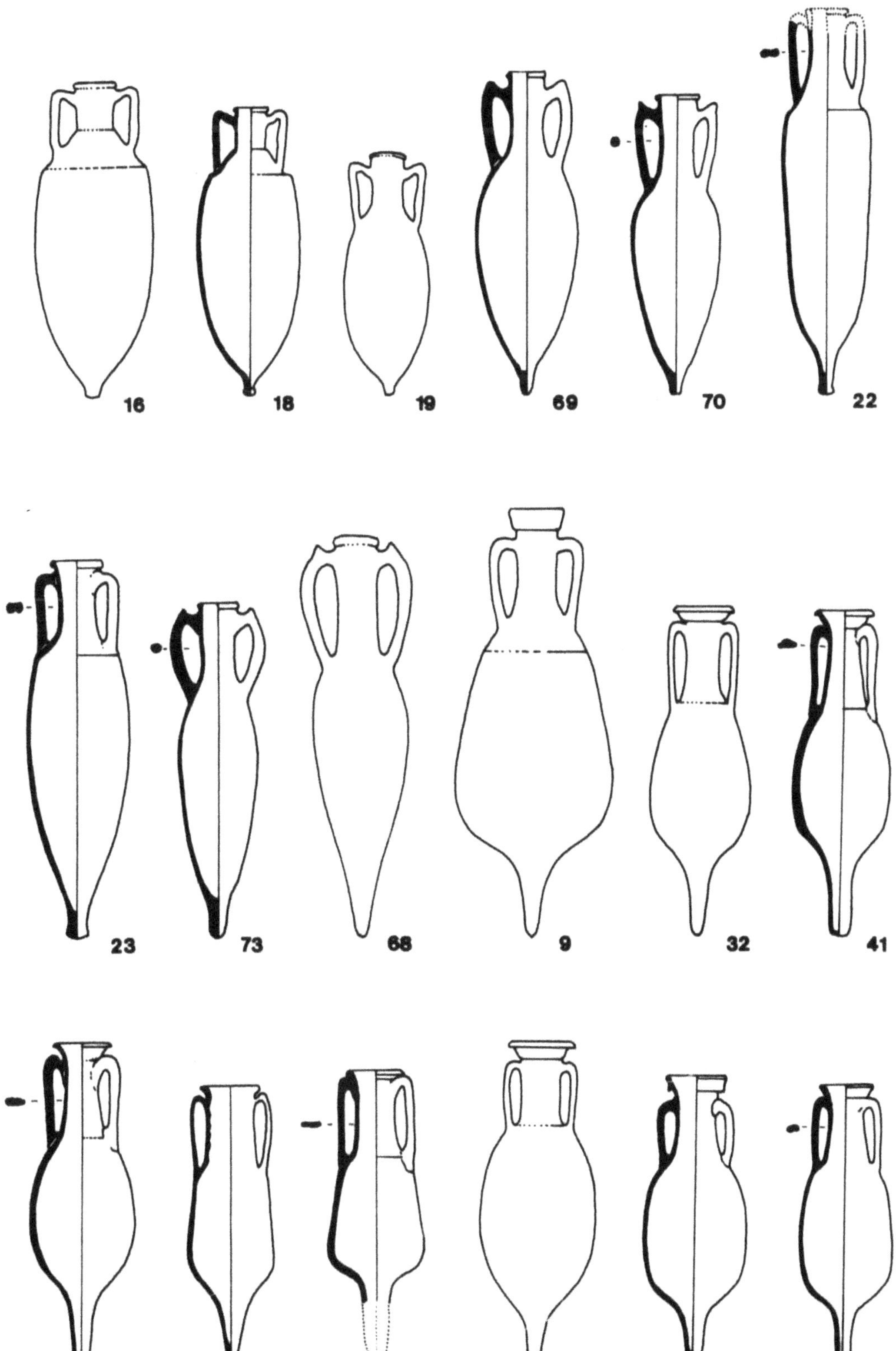

Figure 41. Types of amphorae according to C. Panella

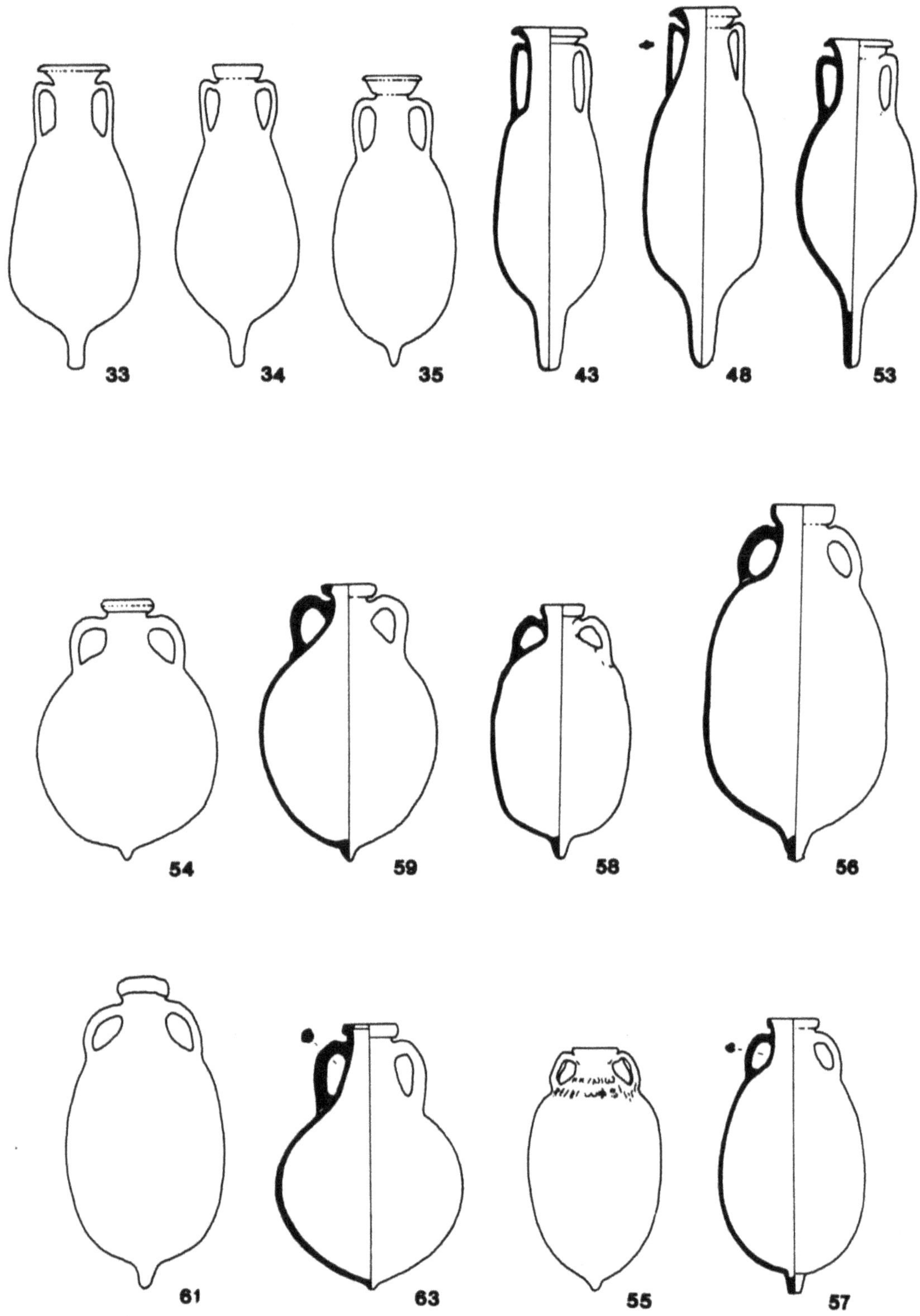

Figure 42. Types of amphorae according to C. Panella

ADATLISTA "GO"

CASE NO.	LABEL	2 LL	3 HL	4 LC	5 HC	6 LPO	7 HH	8 HB	9 LP	10 LF	11 HF	12 FL	13 CD	14 HA	15 AL	16 Z	17 X	18 Y
1	GO01	10.250	8	5.500	35.600	6.500	6.500	63	16	6	22.500	3	30	37	7.500	12	85	11.500
2	GO02	8.500	5.500	5.250	32.500	6	7	52	15.800	6.900	24	2.800	47	35	8	13	67	8.700
3	GO04	10	6.800	5.800	30	5.900	7.500	56	15.500	5.500	21.800	3	38	35	6	11.200	115	7
4	GO05	9.300	3.800	6.500	15.300	7	10	51	20.500	5.500	9	4	27	20.500	6	14	42	5.300
5	GO06	8.800	4.500	6.600	23	7.500	7.500	50	20.800	7	11	4	24.500	27	7	15.500	52	7
6	GO08	10.500	4.200	7.800	22.500	8.500	10	49	18.800	5	12	5.500	31	28	7	14.500	60	6
7	GO09	8.500	6	6	22.300	6.800	13.500	62	24	8.500	23	4.500	28	28	8	11.500	61	8.500
8	GO11	8	2	4.200	19.500	4.200	7	62	14	4	11	4	32.500	22	6	10	48	5
9	GO12	7	1.500	5	16	5	10	53.500	17	4	11.500	4.500	44	21.500	6.900	12.300	37	5.500
10	GO13	6.500	1	5.500	13.500	5.500	11	66	17	4	7	4.500	47	21	7.200	12	39	2.800
11	GO14	7	2	6.500	18	7	10.500	60	18	6	15	2.500	32.500	25	8	13.500	47	3.500
12	GO15	7.500	2.900	5	14	5	10	65	17	4	9	4.200	40	22	8.500	13.500	35.500	4
13	GO16	7	1.600	5.500	12	5.500	12	64	18	2	3	5	48	21	7.500	13	40	2.500
14	GO18	5	1	4.500	11	4.500	8	59	15.800	2.500	5.500	5	43	18	7	10.700	30	2.500
15	GO19	5.500	1	5	15	5	7	44	13.800	4	5.400	3	25	19.500	6.800	10.900	36	4
16	GO20	6.500	2	5.800	10	5.800	9	55	14.500	5	9	3.500	36	18	7	12	36	2
17	GO21	7	2	5.500	13	5.500	9	68	16	4	7	3	45	20.500	6	12	49	3
18	GO22	7	1.500	5.800	18	6.200	10	69	14	4.600	15	4	55.500	26	6	10.500	55	2.500
19	GO23	8	2	5.500	19	6	7	70	15.700	4.200	13	3	50	24.500	6	12	59	2.800
20	GO24	8	2	5	12	5	7	54	15.500	4.500	9	4.900	45	20.500	7	13	41.500	3.500
21	GO25	7.500	2	5	18	5	7	57	14	7	12.500	3.500	45	23	5	11.600	57	2.800
22	GO26	6	1.500	5.500	16	5.500	10	65	16.500	4.500	9	3.700	37	24	7	12	43	3.500
23	GO30	6	2	5	13	5	18	67	16	4	11.200	5	53.500	26.500	7	11.600	42.500	3
24	GO31	10	5	6	20	6.500	28	28	17.500	5	9.500	3.900	30.500	21	5	10.500	52.500	6.500
25	GO32	9.100	4	5	20	6.500	32	17	15.300	7	24	2.500	18	27	5	10	61	5
26	GO33	11	5.500	5.500	12	6.200	41	18	19.600	5.500	12	4	23	12.500	6	12	35	5.800
27	GO34	8	4	6	13	7	37	20	20	7.500	15	3.500	23	19	6.500	11	45	4.500
28	GO35	9	5.500	5	13	5.500	33	27	18	2.500	6	5	31	19	6.500	11	35.500	7
29	GO37	9.200	4.800	6	16.500	7	27	20.500	15.500	5	16.500	4	22	21	5.500	11	51	6.500
30	GO40	8.500	4.200	5.500	18	7	23.500	20.500	14.700	4.100	15	4	23	22	5	10	69	3
31	GO41	9.500	5.500	5	23.500	8.500	22	19	15	8.500	26.500	3	20	28	5.500	11	62	5
32	GO42	9.500	3	8	19.500	8.500	27	12	13.500	6.500	19	1.500	14	25	5	11.800	72	3.500
33	GO43	11.500	5	6.700	20	8	45	23	17	5.500	18.700	3.500	26	23	5	12.500	61	5
34	GO46	9.500	5	5	23	7.600	24	20	16.500	7.500	22	3	22	28	7	10.500	54	5
35	GO47	9	3	6	23	9	27	9	11	6.500	23.500	1.500	11.500	26	3	11.800	127	1.500
36	GO48	11	6	6.300	14	8.500	36	26	18	7.500	24	5	28	30.600	5	11.500	44	6
37	GO53	11	6	7	11	7.500	31	26	18.500	5	33	3	29	19.500	6	12	52	5
38	GO54	8	3.800	6.500	12.100	9	27	30	27	3	4	7.500	38	16	11	17	21	5
39	GO55	7	2.500	6	6	7	28.500	30	20.500	4	4	4.500	33.500	11.500	7	13	22	2
40	GO56	10	5.500	8	15.500	13	44.500	29.500	29	7	10	6.500	36	18	12	19	18	6
41	GO57	8	3	6.800	12	11	31	28	13.500	3.500	6.800	6.500	34	13	8	14.500	26	4
42	GO58	7	3	5	7.500	7	28	30	21.500	4	6	7	34	13	10	15.800	20	4
43	GO59	8.400	4	5.500	12	11	27.500	34	27.500	4	5	6.400	40	19	12.500	18.500	16	4.500
44	GO61	8	5	6	16	12	38	35	24	3.500	8	7.500	40	18.500	12	18	27.500	6
45	GO63	8.500	4	7	18	12	26	28	29	2	2	6	38	21.500	10	16.500	38.500	5
46	GO68	7	2.800	6.200	33	7.500	17	47	13.500	4.500	19	2	48	37	10.500	10.500	90	3
47	GO69	6	1.500	5	22.500	6	22.500	39	15.500	3	8	3	41	26.500	8	10.500	52	4.500
48	GO70	6.500	1.500	5	21.500	6	18	39	14	3	7.500	2.300	41	27	7	10.200	63	4
49	GO73	6	1.800	5.500	17.500	5	2.500	42	12	4	12	2	43	28	8.500	8.500	72	2

Figure 43. Measurements of amphorae in Figures 40–42

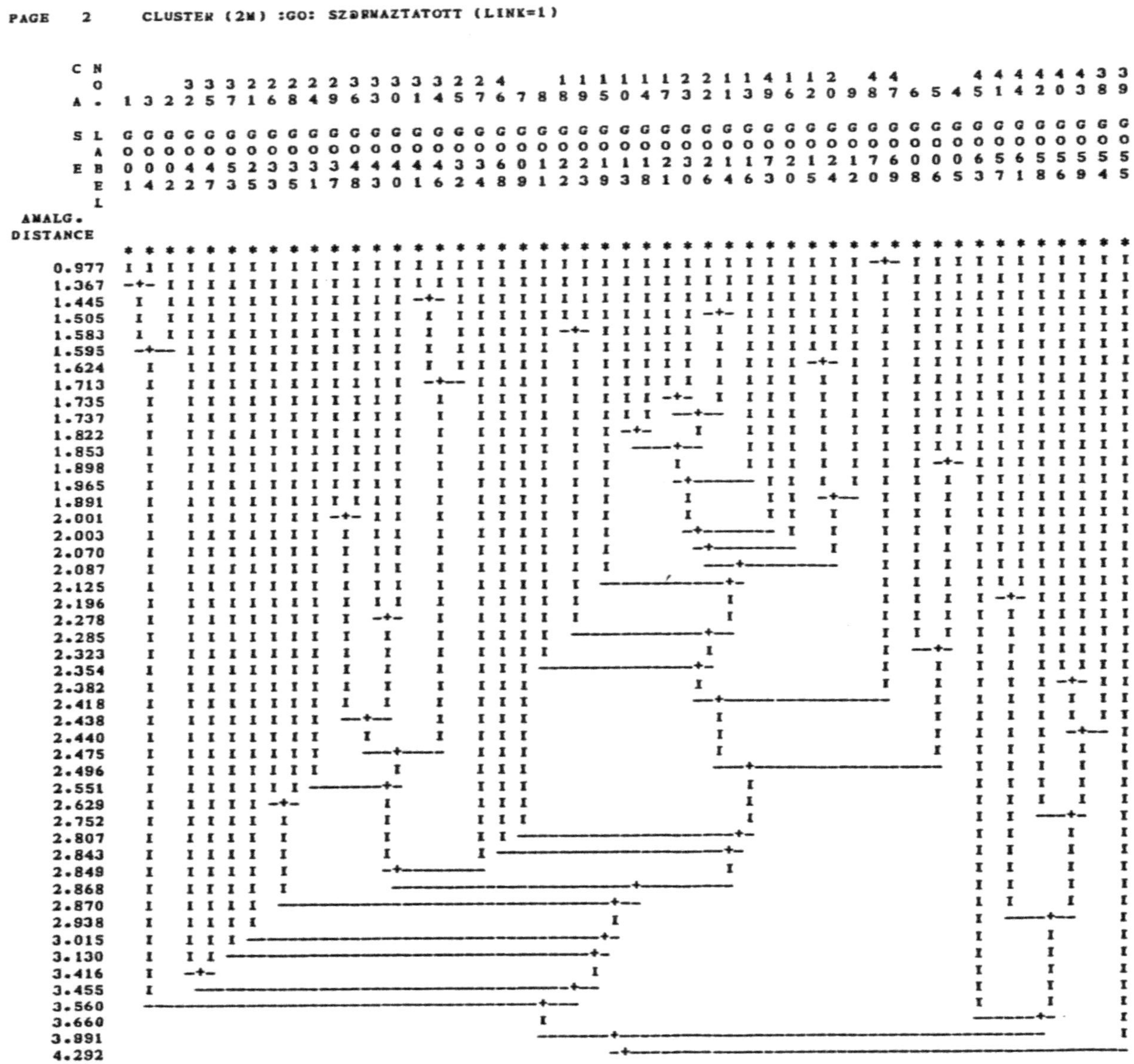

Figure 44. Dendrogram of the amphorae, BMDP Cluster (2M) program (Link = 1)

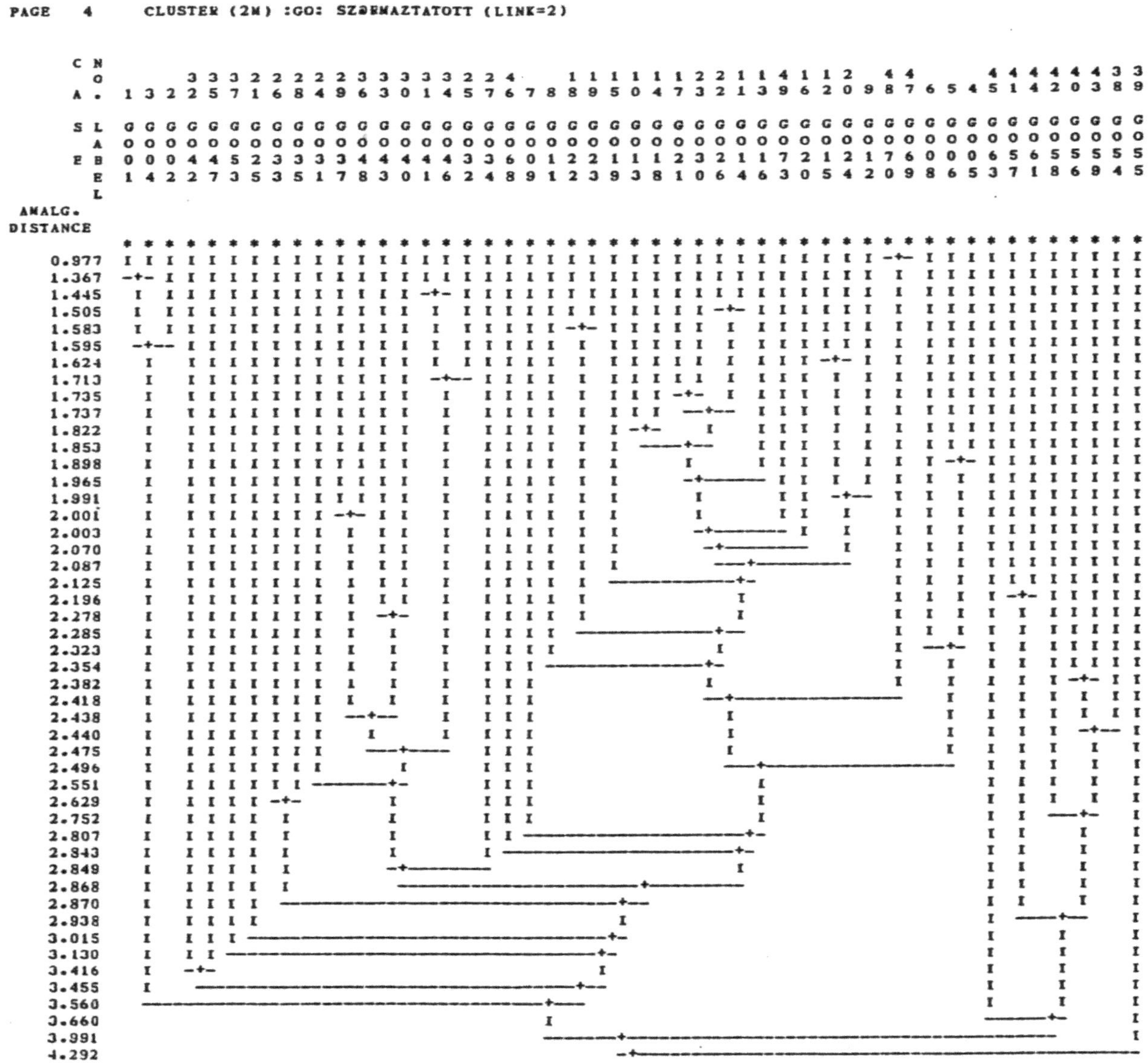

Figure 45. Dendrogram of the amphorae, BMDP Cluster (2M) program (Link = 2)

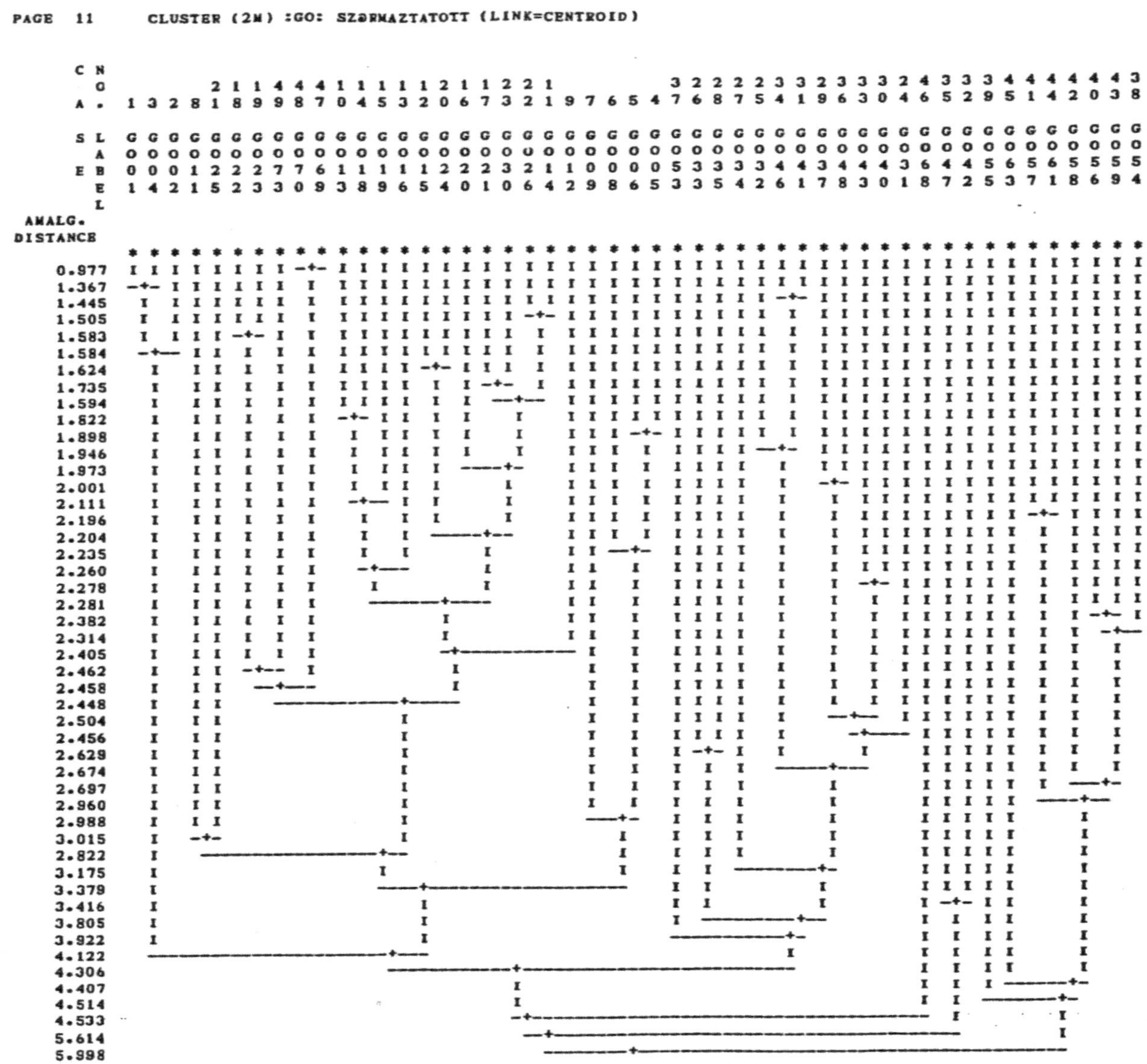

Figure 46. Dendrogram of the amphorae, BMDP Cluster (2M) program (Link = CENTROID)

On the basis of the above numerical characteristics, the amphorae were classified with different computer clustering methods. The classes obtained were compared with typologies commonly accepted by amphora researchers. In that way we tried to make the definition of "similarity" in our model as adequate as possible.

In the following, we publish some results (with no attempt at completeness) in which different parameters were used (figs. 43-46).

What we achieved was mostly the calibration of method, based upon the attribute system of Guénoche and Tchernia. This method gave us an acceptable model of similarity vs. difference, and the analysis of the results demonstrated that the Guénoche - Tchernia system offers an opportunity to make observations within one type only.

More precisely, it neglects the examination of a few characteristic parts.

The extension of the Guénoche - Tchernia method

Our observations have demonstrated that a finer evaluation is required by handle form and mouth profile, as well as by handle cross section (fig. 47).

- The values of the observations related to the rim (19) profile were taken from the table compiled by J.C. Garden.[163]

PROFILE / BENDING	concave v	straight d	convex x
i divergent	vi	di	xi
u parallel	vu	du	xu
o convergent	vo	do	xo

1. xu
2. di or xi
3. xo and xi
4. xo
5. xi

- The observations related to the form of the handle:

a. Is there breakage on the profile? (21)
   1. no
   2. yes
b. Does the profile contain a long vertical line segment? (22)
   1. no
   2. yes
c. Do the upper and lower parts meet in a peak? (23)
   1. no
   2. yes
d. The connection of the upper part and neck is (24)
   1. horizontal
   2. goes upwards
   3. goes downwards

- The profile of the cross section of the handle is (20)

1. flat
2. semi-circular
3. circular
4. divided into two parts

C A S E No.	Label	19 SZÁJ	20 FÜLK	21 FÜLT	22 FÜLEGY	23 FÜLCS	24 CSATL
1	G001	2	1	2	2	1	1
2	G002	2	1	2	2	1	1
3	G004	2	1	2	2	1	2
4	G005	2	1	2	2	1	1
5	G006	2	1	1	2	1	2
6	G008	2	1	1	1	1	2
7	G009	5	1	2	2	1	1
8	G011	4	4	2	2	1	3
9	G012	1	4	2	2	2	3
10	G013	1	4	2	2	1	1
11	G014	1	0	2	2	2	1
12	G015	1	4	2	2	1	3
13	G016	1	4	2	2	1	1
14	G018	1	4	2	2	2	3
15	G019	1	4	2	2	2	3
16	G020	1	4	2	2	2	3
17	G021	1	4	2	2	1	1
18	G022	1	4	2	2	1	3
19	G023	1	4	2	2	1	1
20	G024	1	4	2	2	2	3
21	G025	1	4	2	2	1	2
22	G026	1	0	2	2	2	3
23	G030	1	0	2	2	2	3
24	G031	3	1	2	2	1	1
25	G032	3	1	2	2	1	1
26	G033	3	1	2	1	1	1
27	G034	5	1	2	1	1	3
28	G035	5	1	2	1	1	1
29	G037	2	1	1	1	1	2
30	G040	3	1	1	2	1	3
31	G041	3	1	1	2	1	3
32	G042	3	1	1	2	1	2
33	G043	3	1	1	2	1	3
34	G046	3	1	1	2	1	3
35	G047	3	1	1	2	1	3
36	G048	3	1	1	2	1	3
37	G053	3	1	2	2	1	3
38	G054	1	3	1	1	1	3
39	G055	5	3	1	1	1	3
40	G056	5	3	1	1	1	3
41	G057	5	3	1	1	1	3
42	G058	5	3	1	1	1	3
43	G059	5	3	1	1	1	3
44	G061	1	3	1	1	1	2
45	G063	1	3	1	1	1	3
46	G068	1	3	2	1	2	3
47	G069	1	3	2	1	2	3
48	G070	1	3	2	1	2	3
49	G073	1	3	2	1	2	3

Figure 47 Analysis handle, rim, handle cross section

The extended list of characteristic data is to be found in figure. Naturally, the new items were taken into account during clustering only after an appropriate normalization of the gemoetric data. The description system supplemented in this way has given us satisfactory results. Through its utilization, the clustering method "2M" resulted in an adequate classification from the archaeological point of view. The final results are given in figure 47.

Summary

In this section, an archaeological application of computer statistics methods and results has been presented. Effective, quantitative methods have been elaborated primarily for classifying amphorae according to shape (contour).

The methods - the correct model was selected and the results were tested by comparing them with the typology widely accepted by the research community.

In the course of our work we succeeded in defining an attribute system, grasping the information hidden in the shape of the objects in a more accurate manner than hitherto. Building on this model, we managed to achieve a classification of the samples examined which is adequate from an archaeological point of view.

We believe that the method outlined is the most important outcome of the paper. Its most important element was the selection of the adequate model. We achieved it by evaluating the computer results from archaeological points of view. That methodology may be easily generalized to other groups of objects and considerations of classifications. In the next phase of our work we shall realize an on-line implementation of the above methods based on a personal computer. We also plan to refine our model to include other characteristics (e.g. fabric quality, age, stamps, inscriptions, etc.)

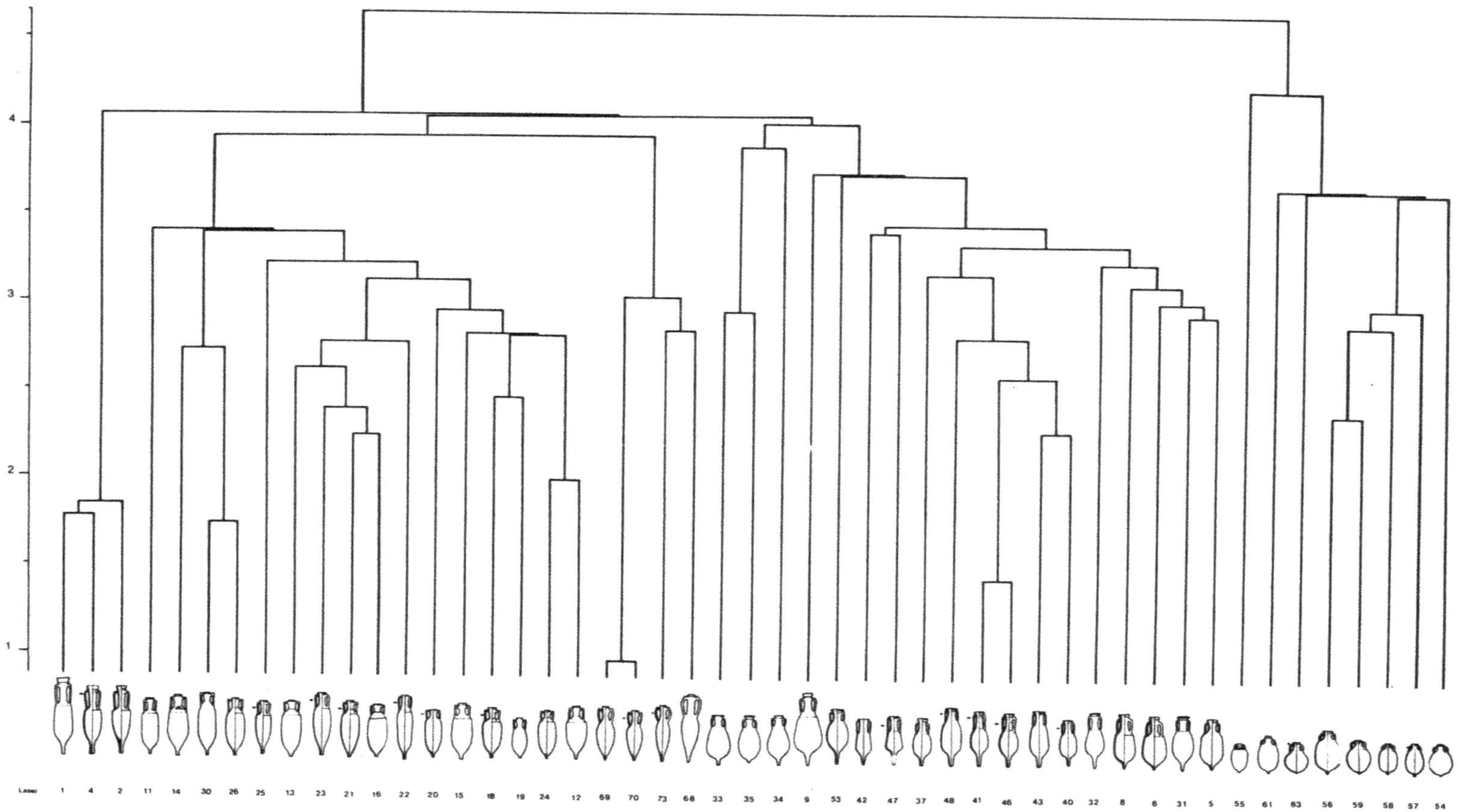

Figure 48. The modified dendrogram of the amphorae

Notes

160. A. Guenoche & A. Tchernia: 'Essai de construction d'un modelle descriptif des amphores Dr. 20'. In Methodes Classiques... (1977), 242-259.

161. cf. note 4.

162. BMDP Statistical Program, University of California, Los Angeles 1982.

163. J.C. Gardin, 'Method for the descriptive analysis of archaeological material' American Antiquity 32.1, 1967, 13-30.

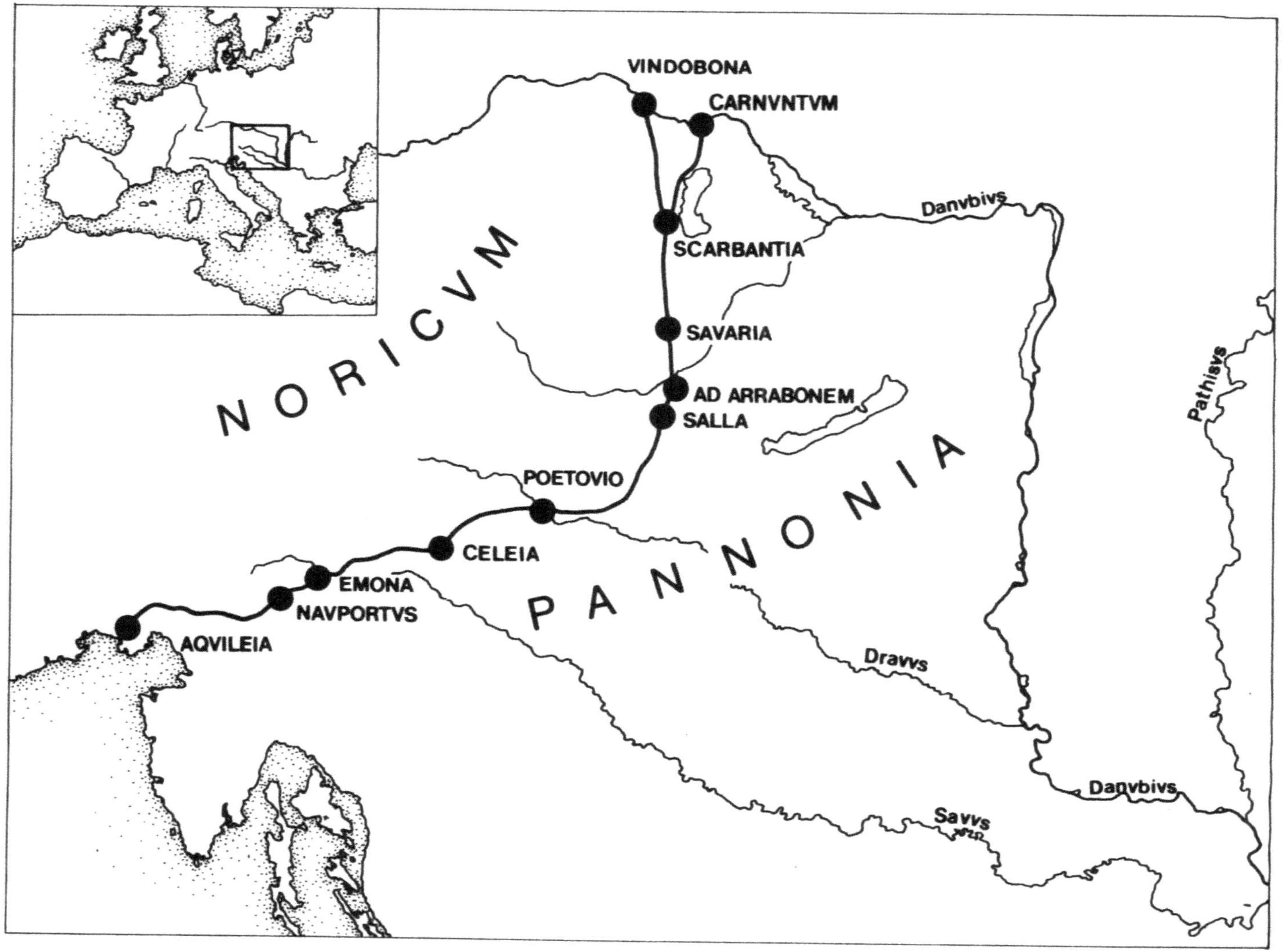

Figure 49. Amber Route sites

www.ingramcontent.com/pod-product-compliance
Lightning Source LLC
LaVergne TN
LVHW071209210926
848482LV00015B/760